BLACK ROBES, WHITE COATS

BLACK ROBES, WHITE COATS

The Puzzle of Judicial Policymaking and Scientific Evidence

Rebecca C. Harris

RUTGERS UNIVERSITY PRESS
NEW BRUNSWICK, NEW JERSEY, AND LONDON

Library of Congress Cataloging-in-Publication Data

Harris, Rebecca C., 1972–
Black robes, white coats : the puzzle of judicial policymaking and scientific evidence / Rebecca C. Harris.
    p.   cm.
Includes bibliographical references and index.
ISBN 978–0-8135–4368–0 (hardcover : alk. paper)—ISBN 978–0-8135–4369–7 (pbk. : alk. paper)
1. Evidence, Expert—United States. 2. Admissible evidence—United States. 3. State courts—United States. 4. Judicial discretion—United States. I. Title.
KF8961.H37 2008
347.73'67—dc22

2008000900

A British Cataloging-in-Publication record for this book is available from the British Library.

Copyright © 2008 by Rebecca C. Harris

All rights reserved

No part of this book may be reproduced or utilized in any form or by any means, electronic or mechanical, or by any information storage and retrieval system, without written permission from the publisher. Please contact Rutgers University Press, 100 Joyce Kilmer Avenue, Piscataway, NJ 08854–8099. The only exception to this prohibition is "fair use" as defined by U.S. copyright law.

Visit our Web site: http://rutgerspress.rutgers.edu

Manufactured in the United States of America

CONTENTS

List of Tables   vii
Introduction   1

1 The Mystery of the Gatekeepers   7
2 Clues to Judicial Behavior   14
3 Forensic DNA: Law Enforcement in the Laboratory   36
4 Lie Detection: Victim of Law and Politics   68
5 Syndrome Evidence: Science Isn't Everything   105
6 Gatekeepers and the Politics of Knowledge   134
7 New Clues? Gatekeeping and the Twenty-first Century   143

Appendix A   State Supreme Court Cases for Forensic DNA   165

Appendix B   State Supreme Court Decisions for Polygraph Evidence   171

Appendix C   State Supreme Court Decisions for Syndrome Evidence   177

Notes   181
Bibliography   183
Index   191

TABLES

Table 3.1	Admissibility of DNA in State Supreme Courts	41
Table 3.2	Decision Patterns for Single-Decision and Dynamic Jurisdictions	44
Table 3.3	Comparing Legal Standards and Judicial Outcomes	47
Table 3.4	Jurisdiction Patterns and Legal Standards for Those Initially Rejecting DNA Evidence	48
Table 3.5	Partisanship and DNA Admissibility	51
Table 3.6	Region and DNA Gatekeeping Outcomes	53
Table 3.7	Types of Experts in DNA Cases	57
Table 3.8	Role of Defense Experts in DNA Cases	57
Table 3.9	Amicus Briefs in DNA Cases	59
Table 3.10	Third-Party Reports and DNA Admissibility	62
Table 3.11	DNA Court Activity Before and After 1996	63
Table 3.12	Logit Analysis of Judicial Outcomes for DNA Evidence	64
Table 4.1	Admissibility of Polygraph Evidence in State Supreme Courts	77
Table 4.2	Holdings on Admissibility of Polygraph Evidence in State Supreme Courts	80
Table 4.3	Jurisdiction Patterns for Dynamic Jurisdictions	83
Table 4.4	Jurisdiction Patterns for Monotonic Jurisdictions	84
Table 4.5	Summary of Judicial Debate	87
Table 4.6	Comparing Legal Standards and Judicial Outcomes for Polygraph Evidence	89

Table 4.7	Comparing Legal Standards and Judicial Holdings for Polygraph Evidence	90
Table 4.8	Partisanship and Judicial Votes to Admit Polygraph Evidence	93
Table 4.9	Partisanship and Judicial Support for Defendants and Prosecutors	93
Table 4.10	Region and Polygraph Gatekeeping Outcomes	94
Table 4.11	Success of Proponents of Polygraph Evidence (Judicial Outcomes)	95
Table 4.12	Success of Proponents of Polygraph Evidence (Judicial Holdings)	96
Table 4.13	Proponents of Stipulated Polygraph Evidence	96
Table 4.14	Logit Analysis of Judicial Outcomes for Polygraph Evidence	100
Table 5.1	Admissibility of Rape Trauma Syndrome Evidence in State Supreme Courts	110
Table 5.2	Admissibility of Battered Woman Syndrome Evidence in State Supreme Courts	111
Table 5.3	Holdings for Syndrome Evidence	112
Table 5.4	Jurisdiction Patterns for Dynamic Jurisdictions	114
Table 5.5	Summary of Judicial Debate for Syndrome Evidence	116
Table 5.6	Legal Standards and Judicial Outcomes for Syndrome Evidence	120
Table 5.7	The Role of Legal Facts in Syndrome Admissibility	121
Table 5.8	Partisanship and Judicial Votes to Admit Syndrome Evidence	123
Table 5.9	Partisanship and Judicial Support for Defendants and Prosecutors in BWS Cases	124
Table 5.10	Partisanship and Dissenting Opinions	124
Table 5.11	Region and Gatekeeping Outcomes for Syndrome Evidence	125
Table 5.12	Success of Proponents of Syndrome Evidence	127
Table 5.13	Success of Proponents of Battered Woman Syndrome	127
Table 5.14	Amicus Briefs in Syndrome Cases	130

BLACK ROBES, WHITE COATS

INTRODUCTION

This book is about a particular interaction between two very different creatures: judges and scientists. These professions outfit themselves in special attire symbolizing the nature of their work. Judges are in the business of processing defendants, using wisdom and legal expertise. The black robe of the judge represents the neutrality and authority requisite for this particular public service. By contrast, scientists are in the business of developing and harnessing the knowledge of how the world works. The white coat of the scientist symbolizes precision and clinical sterility. The precision and clinical sterility of the scientific reputation provides legitimacy to the kind of knowledge arising from a scientific understanding of natural and social phenomena. The interaction of these professions occurs because the white coats often develop and harness knowledge that is very useful to the black robes. The black robes, however, are constrained by political wisdom and legal expectations when they are introduced to a new product from the white coats. They cannot accept every novel scientific advancement as legally permissible in a court of law. Rather, they are required to determine the prudence and acceptability of harnessing a particular science for legal purposes.

The judicial decision to admit or deny a particular science to be included in a legal proceeding is commonly referred to as a gatekeeping decision. Gatekeeping decisions about scientific knowledge abound in everyday judicial process, and the most important gatekeeping decisions are statewide policy decisions by state supreme courts. State supreme courts must decide if novel scientific evidence meets the jurisdictional standard for admissibility in order to certify it for inclusion in a judicial proceeding. Forensic DNA and rape trauma syndrome are examples of scientific knowledge that may arise in judicial proceedings. Two examples of real-world cases from the Supreme Court of Arizona and the Supreme Court of Kansas provide a picture of the way these scientific policy questions are embedded in criminal proceedings. In these two cases, the legal standard for admissibility requires that the science be "found to be generally accepted as reliable in the relevant scientific community" (*U.S. v. Frye* [1923]).

The first example of judicial gatekeeping policy making is from a case arising in Arizona. In 1990 Richard Bible was tried and convicted of first-degree murder. The prosecution successfully introduced evidence of DNA on the defendant's shirt that matched the victim's. The lab had concluded the chances were one in sixty million that the blood was not the victim's. Richard Bible appealed the introduction of this evidence as unreliable. In *Arizona v. Bible*, 858 P.2d 1152 (1993), after extensive discussion and investigation of the reliability of DNA evidence and match statistics, the Supreme Court of Arizona concluded that the trial court erred in admitting DNA probability testimony, as the evidence did not meet the criteria of general acceptance by the scientific community. This case and decision by the Supreme Court of Arizona had the effect of creating DNA admissibility policy for all criminal courts in the state of Arizona.

A second example of judicial gatekeeping policy making is from a case arising in Kansas. In 1980 Elmore Marks Jr. was convicted of raping a woman he met in a club and later enticed to go to a friend's house. At issue was whether the ensuing sexual encounter was consensual. The prosecution introduced expert-witness testimony that the victim was suffering from rape trauma syndrome as evidence that a forcible assault had occurred. Elmore Marks Jr. appealed the introduction of syndrome evidence as unreliable. In *Kansas v. Marks*, 647 P.2d 1292 (1982), the Supreme Court of Kansas upheld the introduction of rape trauma syndrome testimony as "detectable and reliable as evidence" that a rape occurred (654). This had the effect of setting a statewide policy for the admissibility of rape trauma syndrome testimony.

As these brief examples illustrate, the black robes are confronted with the knowledge of the white coats and called on to make a judgment about the validity and reliability of that knowledge in terms of statewide judicial policy. It is easy to understand why this decision-making capacity could be a source of considerable controversy. Judges are not scientists, and their decisions have a very real effect on judicial winners and losers, especially in terms of criminal prosecution. In the examples above, the Arizona case had the effect of shutting down DNA prosecutions until a later state supreme court revisited and reversed the decision. The Kansas case had the opposite effect of providing prosecutors with an additional tool in rape prosecutions.

As a result of this very real policy power, whereby state supreme courts and even individual judges can decide the admissibility of important evidentiary tools, gatekeeping activity has been the subject of much legal and political commentary. Legal scholars have been discussing this particular judicial role in earnest since a 1993 U.S. Supreme Court decision, *Daubert v. Merrill Dow Pharmaceuticals*, 113 S.Ct. 2786 (1993), revisited and rewrote gatekeeping expectations. Here is but a sampling of the gatekeeping conversation appearing in the legal profession:

- "Scientific Panels Should Determine Causation in Mass Tort Cases," *Class Action Law Monitor*, February 15, 2006, 24.

- Tara Marie La Morte, "Sleeping Gatekeepers: *United States v. Llera Plaza* and the Unreliability of Forensic Fingerprinting Evidence Under *Daubert*," *Albany Law Journal of Science and Technology* 171, no. 14 (2003): 171–214.
- "Navigating Uncertainty: Gatekeeping in the Absence of Hard Science," *Harvard Law Review* 1467, no. 113 (April 2000): 1467–1484.
- Lawrence S. Pinsky, "The Use of Scientific Peer Review and Colloquia to Assist Judges in the Admissibility Gatekeeping Mandated by *Daubert*," *Houston Law Review* 527, no. 34 (Summer 1997): 527–578.
- "Improving Judicial Gatekeeping: Technical Advisors and Scientific Evidence," *Harvard Law Review* 941, no. 110 (February 1997): 941–958.
- Amy T. Schutz, "The New Gatekeepers: Judging Scientific Evidence in a Post-*Frye* World," *North Carolina Law Review* 1060, no. 72 (April 1994): 1060–1084.

Many more articles could have been cited, but these few represent the flavor of the debate and discussion. There are several themes articulated in this brief sample. First, scholars disagree about the ability of judges to decide scientific admissibility questions, and this has led to a call for "expert" decision making in the sense of scientific panels, scientific peer review, and technical advisors. Second, there is a sense that this policy problem is new or newly created by recent increases in judicial power (either because of the increased use of sophisticated forensic science or because of changes created by new legal decisions). Third, there is a sense that gatekeeping policy has destabilized. The idea that courts are "navigating uncertainty" or that heretofore-accepted criminal evidence, such as fingerprints, is now questioned, demonstrates that the legal system is in a state of flux and readjustment. Gatekeeping policy is of great concern to those who study and practice law, and it should be of concern to citizens and policy makers.

Legal experts are not the only ones concerned about the gatekeeping power of state supreme courts. Government and law enforcement agencies have likewise demonstrated concern and interest in this particular judicial role. Most recent commentary has been focused on judicial gatekeeping with regard to the most powerful criminal prosecution tool the white coats have produced in decades: forensic DNA evidence. In a 1996 *Medical Science Law* article entitled "Forensic DNA Evidence and the United States Government," two legal scholars detailed the work of all three branches of government to "speed the admissibility of forensic DNA analysis as scientifically acceptable evidence in US Courts." The article goes on to detail the way that new gatekeeping practices with regard to scientific evidence can and will affect the distribution of criminal justice. In more recent developments, the president of the United States released a DNA initiative entitled *Advancing Justice Through DNA Technology* in March 2003. Among other things, the briefing called for increased funding for the scientific education of judges as gatekeepers (*Advancing Justice*, 9).

With so much concern among those carefully watching the activity of the black robes, it seems very timely and necessary to carefully examine the way that the judiciary processes scientific information. Furthermore, gatekeeping activity generates a host of public-policy questions and empirical puzzles: How well do the judges do? What does this mean for science? For justice? How do we explain the way that different courts will reach different conclusions about the same science? To answer these questions and puzzles about the judicial processing of scientific information, the best place to start is with an investigation of gatekeeping activity: who, what, when, where, and why.

This book will begin with a conceptualization of gatekeeping decisions and subsequent empirical and political concerns. The key question of the book is to determine the conditions under which those who wear the black robe recognize and approve of the knowledge developed and harnessed by those who wear the white coats. The puzzle occurs when different jurisdictions reach different conclusions about the *same* science. In other words, while Arizona rejected DNA evidence, Virginia and many other states had admitted it as of 1993. This book will focus on the puzzle of state supreme court variation.

Chapter 1 introduces the "mystery of the gatekeepers" and the empirical puzzles of gatekeeping activity. The mystery of the gatekeepers turns on the ability to explain and predict gatekeeping decisions. Part of the answer to this mystery lies in an understanding of gatekeeping as a political phenomenon—where the constraints and realities of political behavior and the interactions of political institutions are important pieces of the gatekeeping puzzle. Another part of the answer must realize the power of scientific knowledge in the judicial political system. The interaction of politics and science becomes very obvious when two empirical puzzles are considered: the puzzle of state variance and the puzzle of the way different types of science are treated in the judicial system. Chapter 1 also introduces the importance of state supreme court policy making in our political system, and the ways that political scientists have studied state supreme court activity.

Chapter 2 will describe several clues to judicial behavior. These "clues" are competing political science understandings of the explanation of judicial policy outcomes. In order to develop a model of judicial gatekeeping behavior it is necessary to describe the theoretical expectations about factors that are likely to affect gatekeeping decisions. One factor is the law (or legal standard) operating in the jurisdiction. While Arizona and Kansas may use a "general acceptance standard," Idaho or Louisiana may use a relevancy standard or a reliability standard. From a political science perspective it is reasonable to expect that variance in legal standards will lead to variance in judicial outcomes. Another factor is the political persuasion or policy goals of the judges themselves. It may be that Arizona, Kansas, Idaho, and Louisiana have courts with detectable and varied political persuasion. This might mean that the politics of the state or the political preferences of the judges themselves are more supportive of the prosecution than the defendant—or vice versa. From a political science perspective, it would certainly be reasonable to expect courts of different political persuasions to reach different conclusions on scientific admissibility. A third factor is a grouping of several institutional and organizational political variables. For instance, it may be that law enforcement consistently wins admissibility cases. This would imply that state variation is a function of which party (prosecution or defense) is arguing for admissibility. It may also be that there are other institutional factors that explain state variation over time, such as peer court decisions or the presence of third-party reports. Chapter 2 systematically explains these theoretical expectations and concludes with the construction of a model of judicial gatekeeping outcomes suitable for the explanation and prediction of state admissibility decisions.

The remaining chapters test the model through description and analysis of judicial gatekeeping decisions for three types of scientific evidence. Chapter 3 focuses on forensic DNA policy in state supreme courts. An initial timeline of state supreme court decisions shows significant admissibility variation from state to state. Theories about the legal standard at work in the jurisdiction reveal a modest yet significant relationship. Likewise, judicial political variables such as judicial partisanship and regionalism are correlated with judicial gatekeeping policy. Furthermore, important reports from the National Research Council surface as key explanatory factors in judicial outcomes for DNA.

Chapter 4 continues the analysis by looking at the longer timeline of lie detector evidence. The use of polygraphs and their admissibility in court has vexed most jurisdictions. While interesting patterns of law and politics also emerge from this analysis, the chapter compares polygraph gatekeeping jurisprudence with the jurisprudence of forensic DNA. This is important for the testing of the model as well as the explaining of why polygraph gatekeeping policy had a significantly different trajectory than forensic DNA. The chapter tests theories of prosecution success by comparing defendant and prosecution success rates. This analysis is also overlaid with court data to see if particular courts are more prosecution or defendant friendly—possibly explaining state variation in terms of political persuasion.

Chapter 5 takes the analysis one step further by examining gatekeeping policy for two controversial forms of syndrome evidence: rape trauma syndrome and battered woman syndrome. These two syndromes are viewed as a form of science in the judicial system. They are also routinely used by the prosecution (rape trauma syndrome) and the defense (battered woman syndrome). This provides an additional test of the theories explained above: Do legal standards still predict judicial outcomes? Is political persuasion a variable in these admissibility cases? Are there institutional or organizational variables at work here?

Chapter 6 summarizes the findings and discusses the important implications of these observations about gatekeeping behavior. The politics of gatekeeping decisions reveals a significant correlation between law and gatekeeping decisions. The legal standard and the legal facts of particular cases are responsible for a detectable pattern of judicial outcomes. Likewise, partisanship and regionalism are part of the political story surrounding gatekeeping decisions of state supreme courts. Politics also contributes to judicial outcomes through policy advocates (amici briefs), the institutional advantages of the prosecution, and the importance of third-party validation. The theoretical and political implications of these findings are discussed at length, concluding with a discussion of the broader interaction of politics, law, and science in the courtroom.

Chapter 7 returns to the vantage point of the original discussion, recognizing that gatekeeping decisions are public-policy outcomes. With an emphasis on "new clues," this chapter asks if the mystery of the gatekeepers is solved or if there might be additional clues to gatekeeping policy. This final chapter moves the discussion into additional realms of political-science expectation. There are several broad theories of the policy-making process (such as the policy streams approach) that do much to explain and predict judicial outcomes. There are also narrow expectations about additional particularized variables, such as the *CSI* Effect or executive activity, which also augment the above model. This final chapter attempts to illustrate the way that gatekeeping policy is subject to the forces acting on all policies in a democracy. These additional clues to judicial behavior are a starting point for additional empirical work as well as an endpoint for the limits of the explanatory power of the book's model.

CHAPTER 1

THE MYSTERY OF THE GATEKEEPERS

Imagine for a minute that you are observing an ancient society where people live in walled cities and public servants are employed to police the gates. By implicit agreement, these public servants would be charged with orders to allow only certain kinds of strangers into the city—presumably strangers on beneficial business. From your observation point you notice that, indeed, some strangers are being turned away while others are permitted to enter. As you watch various kinds of people approaching the gate, you begin to make a game of it, attempting to guess which strangers will be successful. As you continue observing, you find yourself unable to predict who will gain entry and who will not. You need more information. One thing you recognize, however, is the power of the gatekeepers. As it will be getting dark soon, you have only one question on your mind: under what conditions are strangers permitted to enter the city? You may hypothesize that there are several factors at work: Perhaps the gatekeepers are using some uniform standard to determine who is permitted to enter. Or, perhaps the gatekeepers' attitudes toward particular individuals and their business are at work. Or, it may be that certain strangers have the necessary capital to gain entry into this particular society.

Contemporary American politics has its own form of gatekeepers, especially when it comes to protecting the polity from strangers. I am not talking of human strangers. Rather, I am talking of institutional strangers: particular kinds of knowledge that can create winners and losers in American civil society. In short, the strangers approaching the gates of our polity are novel scientific advancements threatening to redistribute the current allocation of values and resources in the American political system.

The particular locations of interest in this analysis are the gates to the American legal system and the role of judges as gatekeepers. Judges are the policy makers who determine which scientific "strangers" shall be admitted into the halls of justice. As with the observer of the ancient city, the mystery of the gatekeepers is alive and well in the American judicial system. Legal scholars and criminal justice policy experts are unable to predict gatekeeping patterns, and no one has precisely described the conditions under which scientific evidence is validated for judicial purposes.

Gatekeeping as a Political Phenomenon

The secret is found in perceiving gatekeeping as a political phenomenon and using political analysis to understand gatekeepers and gatekeeping. No one would argue the proposition that gatekeepers in an ancient walled city are, in effect, policy makers. Similarly, courts are political actors when they admit or reject scientific evidence. Courts distribute justice by protecting the trier of fact from unreliable or prejudicial scientific evidence. Appellate courts act as gatekeepers for their jurisdictions by allowing some evidence into the courtroom while disallowing other evidence. A gatekeeping decision at the appellate level is both a legislative and a judicial decision. It is legislative because it has the force of law for all jurisdictions in a state. It is judicial because gatekeeping decisions are arrived at by the *interpretation* of rules of admittance. In a sense, this decision signals the proper weapons counsel may use in legal engagement.

Courts allocate political values and resources through common-law decisions. This political power is a source of inquiry for judicial scholars interested in the conditions under which judges make political decisions. In general terms, the dominant question of judicial behavior centers on the determinants of judicial outcomes: why was the outcome x instead of y? As a specific kind of judicial decision, the gatekeeping function of criminal courts presents an interesting context for exploration and testing of theories of judicial outcomes, particularly where the dominant theory—attitudes—offers ambiguous expectations.

Judicial gatekeeping decisions are defined as decisions validating (or invalidating) evidence for judicial use. Novel scientific developments often become public-policy questions in this gatekeeping context, and courts are called on to decide the legal and political value of scientific advances. From a systemic perspective, appellate court decisions often validate certain families of scientific evidence while invalidating other families of scientific evidence. The question of determining conditions under which scientific evidence is validated for judicial purposes is the central question of this book.

The Political Importance of Science

Knowledge is power. The *science* developed and harnessed by the white coats is conceptualized as *a way of knowing*, including theory, such as a theory of deception anxiety, and technical applications of that theory, such as polygraph examination. It has properties of validity and reliability. Obviously the science employed in judicial settings must have legal value. As such, the knowledge developed and harnessed by the white coats can be used by the black robes for two main purposes: identification of key elements in the case (e.g., persons or chemicals) and inference of key causal mechanisms in the case (e.g., the causes of behavior or the causes of damage). Thus, science serves the purpose of assisting the trier of fact with identification and inference.

A few examples will illustrate this function. Many criminal trials involve a question of the identity of the perpetrator. Science has developed several ways of knowing that greatly assist in positive identification. Hair analysis, blood typing, DNA samples, fingerprints, and so forth have all been used by law enforcement to identify perpetrators. Handwriting analysis and voiceprints have also been developed. The inclusion of any of these ways of knowing in a criminal trial could certainly affect the outcome of the trial.

Trial outcomes necessarily create winners and losers, which makes trials political. The way that science is processed in the judicial system provides evidence of its political power. If science plays a critical role in trial outcomes, it can be said to have political power. This political power is particularly evident in the way that state supreme courts carefully craft their gatekeeping decisions. By admitting a particular form of scientific evidence, such as battered woman syndrome, state supreme courts are creating a policy that favors those whose case benefits from the introduction of the evidence. In the case of battered woman syndrome, admitting the evidence favors those women who use it to bolster claims of self-defense when they face murder charges. In a separate example, when state supreme courts reject polygraph (lie detector) evidence, they are favoring defendants, for most lie detector evidence would damage the case of the prosecution. While both of these examples favor defendants, the admissibility of DNA evidence is an example of evidence that generally favors law enforcement and criminal prosecution. As stated above, science has political power because it can be used to influence trial outcomes.

Puzzles in Judicial Gatekeeping

Two empirical puzzles stimulate this research, and it is my contention that these puzzles exist because of the political power of science in the courtroom. Both puzzles involve the variation in judicial treatment of scientific evidence. One puzzle is the differential treatment of the same science from jurisdiction to jurisdiction. It is a puzzle because the science (the way of knowing) is presumably uniform and constant, yet judicial treatment varies from state to state. The other puzzle is the validation of some science and the invalidation of other science. For example, some courts rejected DNA evidence while at the same time accepting rape trauma syndrome evidence. Why is some knowledge admissible while other knowledge is not? One would think "hard" science would be more reliable and valid than science from the psychologist's chair—yet the puzzle of judicial scientific acceptance is there for all to see.

Let us take a closer look at these puzzles. The first empirical puzzle arises from the differential treatment of scientific evidence from jurisdiction to jurisdiction. Why do some states admit a particular type of scientific evidence while others reject it? This sets up a puzzle with regard to the static picture of divergence in jurisdictional behavior. At any given slice in time, some jurisdictions will currently accept the evidence, some jurisdictions will reject the evidence, and some jurisdictions will not be represented. This static picture also creates the second puzzle with regard to the differential treatment of different *types* of science. Why do most states allow battered woman syndrome evidence and forensic DNA while they are starkly divided on rape trauma syndrome and polygraph evidence? This is evidence of divergence in judicial behavior with regard to different types of science. Both the picture of behavior with regard to a particular science and the behavior toward one science versus another create interesting questions for hypothesis testing. How can we account for this empirical picture? Is there any way to explain or predict the way that the black robes will use the white coats?

A third empirical puzzle is the diffusion pattern of judicial acceptance over time. This is a puzzle about the dynamic picture of judicial validation (and invalidation) of scientific evidence. Picture for a moment judicial gatekeeping decisions in various jurisdictions plotted across time. Those decisions (data points) above the timeline would represent validation and those decisions below the line would represent invalidation.

Five patterns would be possible longitudinally across jurisdictions: monotone (straight-line) patterns, shifts or changes in the dominant pattern, indeterminate (scattered) patterns, convergent patterns, and divergent patterns. These patterns create a curiosity about the forces at work in the judicial processing of scientific information. Why are some patterns monotone in favor of acceptance (forensic DNA and battered woman syndrome) while others diverge (polygraph and rape trauma syndrome) with multiple decisions both above and below the line? What accounts for a shift or change toward a particular family of evidence, where the dominant pattern (either above or below) suddenly switches sides and jurisdictions start rejecting (or accepting) where they used to accept (or reject)? More generally, why do some families of evidence experience one pattern while others experience a different pattern?

The question of why shifts occur (or do not occur) is an important one. It presents a second location for hypothesis testing. When these shifts do occur, was it due to doctrinal change, scientific advancement, or political change? The appearance of convergence or divergence similarly produces locations for data testing. What variables account for these differences? Finding the variables responsible for judicial gatekeeping decisions will augment an understanding of the factors responsible for judicial gatekeeping diffusion patterns. The topic of diffusion patterns is revisited in chapter 7.

Why should we try to explain the behavior of the black robes? The central question of judicial outcomes is important for two reasons. Empirically, scholars desire to explain judicial outcomes and perhaps predict future patterns of behavior. Explaining the mechanisms behind differential treatment of scientific evidence could very well allow for the prediction of future outcomes. It also allows for better understanding of the political process and public-policy implications arising from the way the system operates. Theoretically, the differential treatment of scientific evidence presents an opportunity to assess competing models of judicial behavior. In an ironic twist of fate, I myself am a white coat developing and harnessing knowledge about the way black robes process the knowledge developed and harnessed by other white coats. The black robes are the subjects of a research agenda designed to develop and harness knowledge of judicial behavior. Normatively, models of the judicial gatekeeping process allow scholars of politics to evaluate if current practices and realities are acceptable—and, if they are not, where they might be modified to reach the democratic goals of criminal prosecution.

Gatekeeping at the Jurisdictional Level

The empirical context for this work is the study of gatekeeping in American state supreme courts. State supreme courts use common law (court decisions) to formulate and articulate gatekeeping policy for their jurisdiction. Once a state supreme court validates a particular science for legal use, all trial courts in that jurisdiction (or state) must admit the evidence. For example, Virginia Supreme Court was one of the first to admit DNA evidence for criminal prosecution. This had the effect of allowing DNA admissibility in all criminal prosecutions in Virginia. A state supreme court can also decide to invalidate a particular type of science for legal use. When this occurs, trial courts are not permitted to allow the evidence to be admitted during the trial. The study of state courts is a natural source of comparison when we are seeking to understand the causes and effects of particular factors on judicial outcomes. More generally, comparative state research is very useful for testing theories of judicial behavior.

In political science, the study of state supreme courts has taken several directions. First, there have been a variety of descriptive studies on the activities of state supreme courts. Scholars have looked at the determinants of variance in case load (Atkins and Glick 1976). They have also looked at patterns of judicial discretion in the docket (Baum 1977). One scholar even developed a measure of court reputation based on the number of times various state courts were cited by their peers (Caldiera 1985).

The study of state supreme courts, however, predominantly focuses on broadening studies of judicial behavior and increasing understanding about determinants of the judicial vote. One research approach has taken advantage of the comparative institutional settings of state supreme courts to test institutional effects. More specifically, Brace and Hall have done a series of expanding projects on judicial votes in death penalty cases (see generally Brace and Hall 1990, 1993, 1995, 1997; Hall 1987, 1992). Another research approach analyzes the diffusion of policies through court jurisdictions (Glick 1992; Romans 1974). A third approach has studied anomalies in the state courts, especially the low rates of dissent (Sickels 1965) and the factors associated with dissent rates (Jaros and Canon 1971).

The study of state supreme courts has also extended variants of studies of federal courts. Scholars have looked at the role of organized interests in state judicial decisions (Songer and Kuersten 1995; L. Epstein 1994). They have also explored the role of individual attributes on judicial behavior such as the freshman effect (Allen 1991) and religious beliefs (Songer and Tabrizi 1999). A handful of studies have also looked at the factors related to judicial votes on particular policy questions: First Amendment rights (Reid 1988), sex discrimination (Gryski, Main, and Dixon 1986), death penalty (Brace and Hall 1997), and prisoner's rights (Haas 1981, 1982).

The present project is centered on an empirical puzzle: variance in jurisdictional validation of science, rather than variance in the response of individual judges to case stimuli. Here, the decision maker is starkly presented as making decisions under conditions of uncertainty. Of theoretical and political importance is the discovery of the variables that control judicial outcomes. This project takes an ever-broadening approach to truly understand the mechanisms at work in the processing of scientific information by the judicial branch of government.

The present project is also very outcome oriented. The goal is to explain a complex outcome—scientific admissibility in the state courts. It is a natural puzzle because the science is presumably uniform and a constant, yet judicial treatment varies from jurisdiction to jurisdiction. It is my intention to begin to get to the bottom of this mystery and to have something more generally to say about judicial outcomes as political outcomes generated under conditions of uncertainty—especially in a policy frontier where attitudes may be less informative, less multidimensional, or less well formed. At the center of this approach are legal determinants and possibly attitudinal determinants that can control the framing of the question. In other words, the law at work in the jurisdiction as well as the attitudes of the judges are likely partially responsible for state-to-state variation. Broadening out the study is a set of variables that help in evaluation of alternatives—variables that provide leverage within the system, such as the institutional strength of law enforcement or the political influence of interest group amici briefs. The goal of this project is to build on a theory of judicial mechanisms—mechanisms responsible for judicial outcomes. Chapter 2 will explain specific factors in greater detail, yet the questions are very simple: Do patterns of state admissibility vary with law, attitudes, and institutional/organizational leverage? To what extent are these patterns the result of interactions and combinations of factors?

This book joins other studies as a description of state supreme court activity as well as providing a broader theoretical understanding of judicial outcomes. It also develops a foundation for future analysis of diffusion patterns and scientific validation.

DNA, Polygraphs, and Syndrome Evidence

Where to begin? The investigation of judicial gatekeeping requires the selection of case studies, and there are hundreds of specific types of scientific knowledge at use in the legal system. These include typical forensics such as bite-mark analysis, ballistics, hair analysis, and DNA testing. They also include hypnosis, handwriting analysis, voiceprints, and lie detectors as well as psychological tests and indicators of sobriety (Breathalyzers, HGN tests, blood tests). Furthermore, leading authorities such as Westlaw and Giannelli's *Scientific Evidence* also include syndrome evidence (such as rape trauma syndrome and battered woman syndrome) as separate chapters or headings under the general topic of "scientific evidence." With so many ways of knowing at use in the legal system, which types of science provide appropriate case studies for gatekeeping analysis?

DNA evidence, polygraphs, and syndromes were chosen as case studies to address methodological and empirical concerns. First, there was an attempt to include a broad spectrum of science, from the scientist's lab to the psychologist's chair. Second, these sciences have been significantly debated in the judicial setting. Some jurisdictions have accepted these while others have rejected them, providing the variance necessary for testing. Third, these ways of knowing are politically important. The community of legal scholars has generated much discussion about the prudence of admitting DNA, polygraphs, and syndromes into the criminal courtroom.

Empirically, gatekeeping is a relatively narrow judicial function. Therefore it was necessary to choose case studies with the most political importance in order to maximize the meaning of the research. These cases have the largest volume of legal and political discussion, both among court watchers and political actors such as law enforcement and other government agencies. Furthermore, in the criminal setting, these three cases have a significant amount of use and application. To be sure, other controversial sciences existed, such as Horizontal Gaze Nystagamus (HGN), hypnosis, or the application of Luminol treatments, but they were methodologically limited by virtue of being rare. Similarly, other sciences were widespread, such as fingerprints, but lacked significant variation in judicial discussion of validity and reliability. DNA evidence, polygraphs, and syndromes allow for a robust examination of gatekeeping decisions.

Given these empirical parameters, we are ready to advance important hypotheses about the behavior of the black robes. The following chapter will introduce three conventional models of judicial behavior as a starting point for further analysis of DNA, polygraph, and syndrome gatekeeping decisions in state supreme courts.

CHAPTER 2

CLUES TO JUDICIAL BEHAVIOR

The puzzle of state supreme court variance in gatekeeping policy necessitates the search for clues to judicial behavior. When courts from different states are reaching different conclusions about the admissibility of the same science, it is natural to ask for an explanation. State gatekeeping policy is important because of the political implications of admitting or denying scientific evidence in judicial proceedings. For this reason, the analysis centers on *political* explanations of judicial behavior. In fact, a key argument of this book is that judicial gatekeeping is a function of political variables—not scientific variables. Political scientists specialize in recognizing and decoding political patterns in policy outcomes.

With regard to judicial outcomes, political science has developed several competing political theories to explain why different courts reach different decisions. These explanations begin with an understanding of the way human actors (judges and courts) interact in a political system. Part of the story must recognize how humans make decisions. The rest of the story must build on this process by looking at the legal system in which state supreme courts operate. One theoretical approach emphasizes the way the law of admissibility used in a particular state will influence judicial outcomes. Some states simply have stricter legal standards than others, making it more difficult for any science to be admitted in that jurisdiction. A second theoretical approach emphasizes the political attitudes of the policy makers (judges) themselves. Some judges will simply favor law enforcement (or defendants) more than others, as a product of their partisanship, region, and political preferences. For this reason, patterns of admissibility may depend on who is presenting the evidence or who is sitting on the court and where that court is located geographically. A third theoretical approach emphasizes the way the political system provides institutional and organizational advantages for certain political actors. Law enforcement, as an arm of the government, is certainly privileged in numerous important ways in the judicial political system, implying that science introduced by law enforcement should fare better in courts than science introduced by defendants.

This chapter develops these theories step-by-step, with examples of previous use in political science. The goal is to synthesize the approaches to develop a model of judicial gatekeeping that can adequately explain and predict why some state courts were more likely to admit evidence than others. The end of the chapter presents this model as a diagram of related factors. The legal standard (whether strict or lenient) provides the starting point for the court. If the standard is strict, the science will need quite a bit of attitudinal support or political support to overcome the hurdle. If the standard is quite lenient, science can cross the threshold of admissibility with just a little attitudinal or political support. The development of this model precedes chapters presenting the puzzle of state supreme court variance for DNA, polygraphs, and syndrome evidence. In this way, the clues to judicial behavior provide keys to unlock the puzzle of state-to-state policy variance.

Gatekeepers Are Human

As simple as it may sound, it is important, in terms of theory, to begin with the observation that gatekeepers are human beings. The courts developing gatekeeping policy are composed of judges who are human decision makers. Clues to judicial behavior should necessarily be found in clues to the way human beings operate in a decision context. To understand variance in state judicial outcomes, it is important to have an explanation of human decision making to inform our expectations about the way conventional variables will affect judicial outcomes. Because the gatekeepers are human, cognitive psychology is the likely place to begin to look for clues to judicial behavior. It has conceptualized the human decision maker in terms that allow for the synthesis of dominant models of judicial outcomes. Once we articulate the cognitive process involved in decision making, it is easy to see how politically constructed variables, such as legal standards or judicial preferences, may become correlated with state variance in judicial outcomes.

In their study of judicial behavior, Carp and Rowland (1996) have been the most explicit about proposing a theoretical foundation based on cognitive psychology. These scholars recognize cognitive psychology as a science that is sensitive to the way complex human beings operate, and they attempt a synthesis of psychology and judicial studies in order to advance the study of judicial politics. In terms of the puzzle of variance in state-to-state gatekeeping policy, it is important to understand that this differential treatment is the product of human decision makers.

With this in mind, we begin with human cognitive psychology. Carp and Rowland (1996) rely heavily on Herbert Simon's notion of "bounded rationality," a paradigm with strong support in contemporary political science (Carmen 2004, 229–231). Bounded rationality observes that human decision makers have limited cognitive capacity. As such, when they are faced with an ambiguous decision (a decision where probabilities are unknown), they have to rely on "proximate cues" to reach a decision. These "proximate cues" are indicators of the object they wish to perceive but cannot perceive directly. For example, a court confronted with a new scientific method, say the Horizontal Gaze Nystagamus (HGN) test for sobriety, needs to decide if the test is reliable and valid for assessing blood alcohol levels. Law enforcement may employ the test when they stop a motor vehicle, but the judges on the court cannot directly assess the test's reliability, so they must rely on the proximate cues presented by expert testimony in court. Thus, human decision makers are dependent on the provided environment (and their perception of it) when making decisions. Judges also view the "proximate cues" through their own psychological lens—given more weight or credibility to some cues over others. In this way, judges, like all humans, subjectively represent reality and proceed to make decisions based on that reality.

Carp and Rowland point out that small subjective differences in framing can create categorically different outcomes. Clearly, when a human decision maker is weighing alternatives through a framing and evaluation process, we might expect consistent patterns of evaluation from politically constructed information. It is my contention that this is precisely what happens when we see variance in judicial outcomes with regard to scientific evidence. The question and its evaluation have been altered in a predictable manner by particular courts due to particular factors. Thus, the cognitive process, as conceptualized, provides a convincing explanation of the reasons why legal, attitudinal, institutional, and political mechanisms affect judicial outcomes.

Using patterns scholars can detect, certain factors can inform these representations, especially politically defined factors such as law, attitudes, and institutional/organizational variables. In other words, those factors responsible for the proximate clues will affect the gatekeeping decision. To the extent that those factors vary from decision to decision, the outcomes may be expected to vary from decision to decision. Political scientists have articulated three major sets of factors affecting judicial outcomes. First, *law* (doctrine, rules, etc.) is recognized as a constraint on judicial behavior. Judges will consider the law of their jurisdiction when making a decision. Second, political preferences or *attitudes* of individual judges are proven correlates of judicial behavior, and one might expect Republican or Democratic judges to handle cases differently. Third, particularized findings about *institutional and political variables* have demonstrated their correlation with judicial outcomes. Such things as who is presenting the argument, which types of experts are employed, how many attorneys are involved, and so on will also likely affect the outcome of a case.

In the broader context of human decision making, it is easy to conceptualize the way that law, attitudes, institutions, and politics can affect framing and evaluation of scientific evidence for judicial purposes. To put it simply, law, attitudes, institutions, and organizations affect the way an ambiguous question is *framed and evaluated*. These forces operate on the cognitive process to produce the framework used to make the final decision. Law, attitudes, and politics set the values placed on the variables in rational calculation. Furthermore, I believe these are predictable mechanisms that consistently formulate equations for the decision maker.

Gatekeeping and the Law

The first clue to variation in judicial gatekeeping decisions is the presence of variation in the law. Jurisdictions vary in their gatekeeping laws, with some jurisdictions having stricter standards than others. It is necessary to begin by discussing the way law is expected to interact with judicial outcomes generally and gatekeeping decisions specifically. The theoretical approach correlating law with judicial outcomes is known as the *legal model*.

The legal model is the oldest political science model of judicial outcomes. The logic of the model is that "law" determines outcomes. This implies that judicial decision making is a function of stare decisis, original intent, balancing, and plain meaning of the text. In short, judges use legal rules to interpret the significance of the question at hand. This model conceptualizes these rules as a very real constraint on judicial outcomes. One articulation of the legal model (Levi 1949 as cited in George and Epstein 1992) presents the basic pattern of this model of decision making. Here, the court is idealized as using a structure to interpret the law. This involves a step-by-step logical process that is the classical understanding of judicial behavior: the rule is articulated, the facts are articulated, and the rule is applied to the facts to obtain a result. For our purposes, the rules may vary from jurisdiction to jurisdiction—even when the facts are the same. This would likely explain some of the variance between courts making decisions about the same science.

Here, *law* is conceptualized as the *algorithms and calculi* used to determine the validity and reliability of a certain way of knowing for judicial purposes. These are generally conceptualized as doctrines, principles, tests, and methods of reasoning. It is important to note that variation in algorithms may produce variation in judicial determinations; employing different calculi may yield different conclusions with regard to the same science in question. In this way, the employment of a specific calculus may be dispositive of the outcome. At the least, *law* acts as constraint that varies from jurisdiction to jurisdiction. All of this implies that the gatekeeping outcomes of various jurisdictions may have a *legal* pattern.

Scholars (Prichett 1948; Schubert 1960, 1965; Murphy 1964; Segal 1984, 1986; George and Epstein 1992) have examined these "legal" factors to explain judicial outcomes. Indeed, empirical results indicate a substantial amount of support for this "calculus" explanation of judicial outcomes. Segal (1984, 1986) found a significant pattern between doctrine and facts in search and seizure jurisprudence in the Supreme Court. George and Epstein (1992) similarly found a significant pattern between doctrine and facts in death penalty cases. Brace and Hall (1993, 1995, 1997) also found an independent effect for legal factors, even after controlling for attitudes and institutions. Segal (1984, 1986) systematically demonstrated how law controls outcomes based on the six exceptions to warrant cases. For any given case, discrete facts and other legal stimuli significantly influence judicial choice (Brace and Hall 1993; Emmert 1992; Emmert and Traut 1994; George and Epstein 1992; Hall and Brace 1994, 1996; Segal 1984, 1986; Segal and Spaeth 1993; Songer and Haire 1992). These findings clearly imply a role for "legal mechanisms" in judicial outcomes.

The legal model dominates the literature of legal scholars (e.g., lawyers) who discuss the judicial processing of scientific information. These scholars consistently discuss gatekeeping decisions in terms of relevant legal standards and the implications of changes in the common law. Legalists generally feel that the old doctrine will anticipate the new one. In the context of variance in gatekeeping decisions, this model would hypothesize that legal variance (doctrines, precedents, and principles) is correlated with the outcome variance. If this is true, we would expect to find the following outcomes in scientific gatekeeping decisions: (1) doctrinal variance should correlate with variance in outcomes; (2) doctrine should consistently include some science while excluding other science; (3) the role of legal reasoning implies that the same line of reasoning (especially in terms of legal analogy—that is, using fingerprints to explain DNA) should lead to the same outcome; (4) the role of precedent implies that courts will follow earlier findings unless a significant change in doctrine or science occurs to alter the interaction of law and facts.

Thus, as a first-cut analysis, legal variance must be examined as a possible explanatory variable. Legal variance may also account for changes in outcomes over time. According to the legal model, a shift in judicial outcomes should correspond with a change in the legal standard, doctrine, reasoning, or principles applied by judicial personnel. For example, in 1993 a Supreme Court decision (*Daubert v. Merrill Dow Pharmaceuticals*, 113 S.Ct. 2786 [1993]) (subsequently adopted by several state jurisdictions) significantly changed the legal calculus applied to questions of admitting novel scientific evidence. The legal model implies that this should produce a change in judicial outcomes.

How do we decide if a jurisdiction is using a strict or a lenient standard for scientific admissibility? Specification of the legal model is not too difficult in the gatekeeping context. There are essentially three legal standards or doctrines for admitting scientific evidence (Giannelli and Imwinkelried 1993). Some jurisdictions use one standard, while other jurisdictions use another. The first standard is the relevancy standard (known to lawyers as Federal Rule of Evidence 702). The relevancy standard involves a three-step analysis: (1) ascertaining the probative value of evidence, (2) identifying any countervailing considerations in the interest of justice, and (3) balancing the probative value against the identified dangers. Supporters of this standard argue that expert evidence should be admitted as long as the expert was qualified and the evidence would not mislead or prejudice the jury. For many states this is their only admissibility standard.

A second legal doctrine used by most states is known as the *Frye* standard (*Frye v. United States*, 293 F. 1013 [1923]). The *Frye* standard (also referred to as the "general acceptance test") is a rule established in 1923 by the Supreme Court and subsequently adopted by several states. This standard requires all the above as well as requiring "general acceptance in the particular field in which it [the evidence] belongs" (*Frye*, 1014). This doctrine requires the decision maker to define the field of experts and then assess the degree to which they support the reliability and validity of the science at hand.

The third doctrine is called the *Daubert* standard (*Daubert v. Merrill Dow Pharmaceuticals*, 113 S.Ct. 2786 [1993]). The *Daubert* standard is the most recently articulated doctrine of scientific acceptance, arising from a 1993 decision by the Supreme Court. In that decision, the Court said the "general acceptance" test was only one consideration among many. According to the Court, the real question required of the Federal Rule of Evidence (FRE) 702 is that scientific evidence be found "reliable."[1] The Court defined this as "grounded in the methods and procedures of science . . . derived by the scientific method." Then the Court went to list four factors that may assist the gatekeeper in determining reliability: "(1) whether the theory or technique can be (and has been) tested according to the scientific method; (2) whether it has been subjected to peer review and publication; (3) the known or potential rate of error; (4) whether it has been generally accepted" (Bander 1997).

Daubert has been characterized as providing a "wider gate" for scientific evidence even as it calls for a "more vigilant gatekeeper" (Davis 1994). Thus, *Daubert* represents a significant doctrinal departure from *Frye*. Indeed, the first federal appellate case dealing with polygraph evidence post-*Daubert* (*United States v. Posado*, 57 F.3d. 428 [5th Cir. 1995]) indicated that the polygraph, while excluded because of general acceptance, might be accepted with the other criteria. "*Posado* clearly and repeatedly states that a *per se* rule against the admissibility of polygraph evidence is no longer viable after *Daubert*" (Bander 1997, 700). The *Daubert* standard has subsequently been adopted by several state jurisdictions. This is the most complex and complete treatment the U.S. Supreme Court has given scientific evidence, and several states have adopted this standard for their own jurisdiction.

Given this specification, how should law affect judicial gatekeeping decisions? First, different laws or legal standards should correlate with different gatekeeping outcomes. Furthermore, this legal pattern should occur in a predictable manner. Very strict admissibility standards should decrease the admissibility of scientific evidence. Very liberal admissibility standards should increase the admissibility of scientific evidence. In other words, jurisdictions with strict legal standards will be less likely to admit novel scientific evidence, while jurisdictions with loose legal standards will be more likely to admit novel scientific evidence. As the following chapters will demonstrate, a legal pattern for gatekeeping decisions is detectable from jurisdiction to jurisdiction. Indeed, legal standards explain much of the variance in judicial decisions with regard to DNA, polygraphs, and syndrome evidence.

Gatekeeping and Judicial Preferences

A second clue to judicial gatekeeping decisions lies in the political preferences of the judges themselves. Jurisdictions certainly vary in their judicial personnel, and those personnel may have different political goals. The idea that different people are making the gatekeeping decision is, at its root, a story about the way the preferences of those people may affect the outcome of the decision. It assumes that judges have preferences or attitudes about the case before them. The theoretical approach correlating judicial preferences with judicial outcomes is known as the *attitudinal model*.

The logic of the attitudinal model is that the primary goals of judges are policy goals. It posits that judicial decisions are a function of the facts of the case and the attitudes of the justices. Because a judge will have beliefs of a political nature regarding the stimuli in his or her courtroom, this model implies that judicial decisions can be predicted from the justice's political attitudes and personal attributes. The logic of the model is that judges choose the alternative that is closest to their preferred political outcome. For instance, a conservative judge might prefer stronger sentences for convicted criminals. A liberal judge might prefer to have police follow strict rules for gathering and presenting evidence in order to protect the innocent. Data often demonstrate that aggregate judicial behavior reflects attitudinal behavior, with conservative decisions flowing from conservative courts and liberal decisions flowing from liberal courts.

Empirical studies have demonstrated substantial support for the attitudinal model in policy areas easily defined by ideological expectations. There is ample empirical support for the hypothesis that the judges' ideological predispositions (Danelski 1966; Prichett 1941, 1948; Rhode 1972; Rhode and Spaeth 1976; Schubert 1965, 1974; Segal and Cover 1989; Segal and Spaeth 1993) or the personal attributes that serve as their surrogates (Brace and Hall 1993, 1995; Goldman 1966, 1975; Hall and Brace 1992, 1994, 1996; Tate 1981; Tate and Handberg 1991; Ulmer 1970, 1973a) play a crucial role in judicial decision making. Scholars have found attitudinal correlations with the adoption of innovative policies, but there is no evidence of innovativeness as a consistent trait among judges (Baum and Canon 1981).

For purposes of this project, this model suggests the need to test for ideological correlations with judicial outcomes. It would hypothesize that an ideological pattern may fit the pattern of judicial policy making. In other words, it is very possible that an attitudinal pattern is present in court gatekeeping activity. When it comes to specific types of evidence, liberal courts may consistently rule in a direction opposite that of conservative courts.

When it comes to questions of scientific gatekeeping, specification of the attitudinal model is problematic but not impossible. The general hypothesis contends that the political attitudes of judges will be correlated with judicial outcomes. Yet, specifying how a particular court will rule is difficult for several reasons. First, some decisions will have more political meaning than other decisions. For example, if rape trauma syndrome was framed as pro-feminist, attitudes toward feminism and liberalism in general may be activated. By contrast, little political information may be involved in the framing process for a decision about HGN and intoxication. Second, several attitudinal dimensions could influence judicial preferences in a particular case. Attitudes toward the specific parties, attitudes toward perceived political winners/losers, attitudes toward "law-and-order" public policy (e.g., pro-accused or pro-victim), and attitudes toward innovation and change (with regard to both science and judicial roles in public-policy change) are all possible attitudinal dimensions. There is no a priori way to specify which dimension will be activated. Third, it is possible that there are different dimensions at work for different types of science, depending on the framing effect. For example, the law-and-order dimension may dominate attitudes toward DNA evidence, and the innovation dimension may dominate attitudes toward "syndromes." Thus, policy preferences and attitudes might play a role in gatekeeping decisions, but the dimension and the direction is unclear. This does not, however, preclude us from labeling courts and comparing ideology and outcomes.

Ideally, one would like to measure attitudes and policy preferences by surveying judges. Judicial personnel, however, are very reluctant to disclose political preferences because of the perceived role of judges as apolitical and the norm of neutrality (Songer, Sheehan, and Haire 1999). Indeed, many surveys demonstrate that judges do not feel that their political preferences play a conscious role in their decision making. This reality has forced attitudinal scholars to turn to independent observation and surrogate measures to capture an attitudinal effect. Segal and Cover (1989) used newspaper editorials to label judges. Carp and Rowland (1983) used political preferences of appointing executives to label judges (assuming executives appoint judicial personnel with shared values). Songer, Sheehan, and Haire (2000) used party identification. Segal and Spaeth (1993, 2002) used past voting behavior.

Several investigations support using partisanship as a indicator of judicial ideology on courts (Goldman 1966, 1975; Tate 1981; Carp and Rowland 1983; Songer and Davis 1990), including the Supreme Court (Ulmer 1973; Tate 1981; Tate and Handberg 1991), federal district courts (Carp and Rowland 1983; Rowland, Songer, and Carp 1988; Dudley 1989), state appellate courts (Nagel 1961; Ulmer 1962; Hall and Brace 1989, 1992), and U.S. courts of appeals (Goldman 1966, 1975; Songer 1982; Gottschall 1983; Tomasi and Velona 1987; Songer and Davis 1990; Songer and Haire 1992; Songer, Davis, and Haire 1994). Individual judicial votes might correlate with individual judicial party affiliation and thereby allow for the inference that ideological attitudes are related to judicial behavior. Recently, more sophisticated preference measurement has taken the place of "party-adjusted" surrogates (Brace, Langer, and Hall 2000) that link partisanship and state (constituent) ideology.

In terms of initial investigation, it makes sense to compare partisanship with outcomes in gatekeeping decisions. Partisanship is best measured at the individual level, where particular judges are labeled "Republican" or "Democrat." This is fairly easy for many state supreme court justices who run in partisan campaigns. Attitudinal influence in gatekeeping decisions can be inferred if we see Republican judges and Democratic judges voting differently in gatekeeping cases.

Because the unit of analysis here is a court decision rather than an individual decision, it is also important to develop a second way to capture hypothesized attitudinal relationships to outcomes. Individual attributes (e.g., partisanship) often used as surrogates for attitudes are not appropriate for comparing court to court in multivariate analysis. Rather, it is also necessary to develop a way to measure attitudes as they may vary from court to court. One surrogate for attitudes is that of *region* (see generally Songer and Davis 1990). While courts (as a whole) cannot be labeled with a partisan surrogate, they are certainly operating in a single region. Regional differences in judicial decisions have been noted by many scholars and these differences have been attributed to attitudinal differences at work in certain regions of the country. For instance, Freely (1989) found regional differences among judges on state supreme courts, particularly noting that courts in the South behaved differently than other courts in a particular issue area. Other studies found regional effects working on U.S. district and appellate courts (Rowland and Carp 1996; Songer and Davis 1990; Wenner and Dutter 1989). Caldiera (1985) argued that courts in a certain region share similar cultures and values. Thus, while not a perfect measure, the inclusion of region will allow for at least some exploration of potential attitudinal differences in gatekeeping decisions among different courts.

Given this specification, how should judicial preferences affect judicial gatekeeping decisions? First, different attitudinal indicators, such as partisanship or region, should correlate with different gatekeeping outcomes. Furthermore, this attitudinal pattern should occur in a predictable or logical manner. Republican or Democratic treatment of gatekeeping decisions should fit within expectations of partisan preferences with regard to the political context of the gatekeeping decision. In other words, if a gatekeeping decision favors law enforcement at the expense of the defendant, we may expect Republican judges to be more likely than Democrats to admit the evidence. Regional treatment of gatekeeping decisions should likewise reflect known regional differences in criminal justice administration. For example, if Southern states are known to be tough on defendants, we might expect Southern jurisdictions to be more likely than other jurisdictions to admit scientific evidence that is pro-prosecution. As the following chapters will demonstrate, there is some evidence of partisan and regional patterns in gatekeeping jurisprudence.

Gatekeeping and Political Leverage

Further clues to judicial behavior arise from the fact that scientific evidence does not arrive at the courtroom in a political vacuum. Rather, political forces have been at work in several dimensions along the way, and these forces contribute to the way that science is going to be received and processed in the judicial system.

Science comes to the courtroom with a degree of political leverage depending on the political context of the issue, the political strength of the litigants, and a host of other factors listed below. These factors provide scientific evidence with a measure of political leverage. In other words, different science in different cases will vary in the amount of political (institutional and organizational) support necessary to be successfully admitted to the legal system. I refer to this model as the *political leverage model*.

A story about judicial outcomes must recognize that the influence of law and attitudes is mediated by this third set of variables. The legal model cannot independently account for the acceptance or rejection of scientific evidence. Rather, it is subject to institutional and organizational forces that inform the decision maker with respect to the doctrinal "fit" of the stimuli in question. In the gatekeeping context, this implies that variables outside the law may influence the degree to which a particular kind of science meets doctrinal requirements. For instance, strong support by law enforcement and government will provide resources for research and development. This was the case with DNA and with polygraph evidence—both heavily endorsed and utilized by law enforcement. As a result, research and development helped these sciences gain general acceptance in the relevant scientific community. Support by law enforcement may also mean stronger expert testimony and record development in particular cases—thus allowing a particular state court to reach the conclusion DNA or polygraphs are reliable enough for legal use.

Similarly, the attitudinal model is not an independent model. Political preferences of particular courts are also subject to manipulation by institutional and organizational forces, particularly with regard to the judicial frontiers of scientific evidence. While it is conceded that the attitudinal model often predicts behavior on issues that have been ideologically defined outside the judicial sphere (such as reproductive rights), it is not clear how attitudes influence judicial preferences in an ideologically undefined policy setting (such as the admittance of novel scientific evidence in a criminal proceeding). When courts are faced with an ideological frontier, institutional and organizational forces have a role to play in attitude formation. They may inform judicial personnel of ideologically meaningful consequences of judicial outcomes. For instance, a novel scientific approach, such as battered woman syndrome, may greatly help defendants in domestic murder cases. Courts have to decide if it is good conservative or liberal policy to allow battered woman testimony in domestic murder cases. If particular conservative or liberal groups favor or disfavor the evidence, courts may act accordingly—depending on their ideological preference. In this way, information on the attitudinal "fit" of scientific evidence is rooted in a political process, and particular evidence in particular cases arrives at the courtroom with varying amounts of political leverage.

Thus, the third type of variation in gatekeeping decisions is variation in the political factors that may contribute to judicial decisions. The political leverage model is a combination of particular theories citing the role of institutional and political variables in cognitive processing and judicial outcomes. These variables may be empirically associated with judicial outcomes to the extent that they increase or decrease the perception that science meets doctrinal or ideological criteria. In other words, as Carp and Rowland (1996) pointed out earlier, the judge is a decision maker faced with an ambiguous question. He or she cannot *know* if the science is valid and reliable; he or she must rely on proximate cues to gauge reliability and validity. The judge also has to assess competing cues. From the perspective of the cognitive process and from the perspective of a system where institutional and political positioning can help or hinder movement through a system, variables arise. These variables will reduce uncertainty on the part of the decision maker and provide leverage within the system (by virtue of skills, resources, and position).

Political leverage can originate in two places, *institutional structure* and *organizational strength*. Each of these captures several important factors in the judicial process. We will turn our attention to institutional factors first. In terms of institutional structure, important institutions in the judicial system become a source of leverage for certain litigants. Specifically, institutional political leverage may arise from repeat-player status, appellant status, and the policy context generated by peer court decisions to date.

Repeat Players and Litigant Status

Repeat-player status hypothesizes that a certain group of litigants will possess an institutional advantage because they are "repeat players" in the system. Scholars have articulated this relationship in several ways. Galanter (1974) described this status as arising because of (1) association with a national litigating organization or (2) regular appearance before the court. Vose (1959) recognized that attorneys from public interest firms are ensured some level of financial and other support that private counsel may lack, thus providing them with resources and repeat experience. O'Connor (1980) observed that repeated and continuous use of legal recourse breeds familiarity, confidence, and expertise that one-shotters cannot acquire. Segal (1988) noticed the possibility of presumed expertise and deference regarding the solicitor general. Haire, Lindquist, and Hartley (1999) specifically argued that party strength may be viewed in relationship to resources, and repeat-player status before the courts creates familiarity with the institutional practices of particular courts and the decision-making practices of the judges sitting on those courts. Repeat players may also influence the development of legal doctrines in a manner that furthers their own interests. This last group of scholars conceptualized this expertise as containing both a process expertise and a substantive expertise.

Empirical support for this hypothesis has been generated by research in both criminal justice and civil suits (see generally George and Epstein 1992 as well as Galanter 1974; Wheeler et al. 1987; Sheehan, Mishler, and Songer 1992; and Songer and Sheehan 1992). These studies provide some empirical verification for the theory that "haves," including government and corporate litigants, are more likely to prevail because of their enhanced resources, their ability to settle unfavorable cases, their institutional farsightedness, and their repeat-player status, especially when these parties are paired against less powerful individual litigants (but see Sheehan, Mishler, and Songer 1992). Wheeler called them "stronger" and "weaker" parties to give an indication of their likely success in judicial proceedings.

In terms of the processing of scientific information in the criminal justice system, this body of evidence shows that certain litigants are expected to experience increased probabilities of success. For instance, we might expect prosecutors to have more experience with novel scientific evidence than defense attorneys. In terms of cognitive processing, repeat players presumably trigger habitual frames, and courts may come to view prosecutors or defense attorneys differently. Similarly, evaluation of the argument of repeat players presumably carries more weight because of trust and reliability. This implies that we should expect overwhelming evidence of decisions in favor of law enforcement, the most dominant repeat player in criminal law. Yet, do we see it? What does the scientific evidence data say about decisions favoring law enforcement? Furthermore, what can scientific acceptance tell us about the mechanisms influencing judicial decision making—especially under conditions of increased uncertainty. It seems appropriate to look for patterns supporting law enforcement and legal expertise.

Given this specification, how should repeat-player and litigant status affect judicial gatekeeping decisions? Science supported by law enforcement (a repeat player) should be more likely to be accepted than rejected. In other words, when the prosecutor is arguing for admissibility, the gatekeepers should be more persuaded than when the defense is arguing for admissibility. Likewise, when the prosecutor is arguing against admissibility, the gatekeepers should be more persuaded than when the defense is arguing against admissibility.

Appellant Status

The theory of the likelihood that appellants will win is very simple. Logically, we would expect parties to calculate their odds of winning and only expend resources to challenge (appeal) those cases they felt they could win. Thus, judicial outcomes are more likely to favor appellants because of selection bias in the cases tried before appellate courts. George and Epstein (1992) found evidence of this relationship in their work on search-and-seizure cases. Political science scholars anticipate a pattern where appellants are more likely to win. In judicial processing of scientific information, the theoretical question of importance is who is leading the appeal. Are these cases generally ones where science is being challenged on appeal, or where science is trying to prove itself? Which side, pro- or anti-science, has rationally calculated a chance at success? Under what conditions is science likely to win? How often is science successfully defended on appeal? Science supported by appellants is more likely to be accepted than rejected, and for this reason we might expect state supreme court gatekeeping decisions to favor appellants. The only caveat to this expectation is the prevalence of routine appeals in certain criminal cases, particularly capital cases (where DNA evidence, polygraphs, or syndrome evidence might become an issue). For this reason, we will approach appellant status cautiously, recognizing that some defendants will automatically appeal—regardless of whether their odds of winning are high.

Peer Court/Superior Court Decisions

Peer court context is an important proximate cue for judges faced with decisions of uncertainty. Either by emulation of other courts or because of diffusion of judicial policy from court to court, peer court context is clearly expected to influence judicial outcomes. Kilwein and Brisbin (1997) argue that courts emulate "exemplar" or "role model" courts. These scholars point out emulations are not a blind copycat of another court's behavior. Rather they are evidence of learning from the experience of others and applying those lessons to their own decision context. Bennett (1988, 1991) observed that this practice implies a pattern of convergence when innovative policy decisions are working their way through the judicial system. Shapiro (1970) observed this convergence and argued that this is because state supreme courts constantly cite the decisions of others as persuasive, illustrative, or worthy of consideration in making their own rulings.

Clearly, this expectation is reasonable for gatekeeping decisions about novel scientific evidence. Courts are uncertain. They will want to build on the experiences and observations of the courts that tackled the problem in a prior time. They will also be likely to rule in the direction of prior courts. Baum and Canon (1981) observe other patterns in judicial innovation. Of primary concern is the fact that courts are dependent on litigants to rule on a question of law. This can make diffusion of policy in the judicial process "idiosyncratic" but fairly uniform.

Furthermore, there is the idea that the collective wisdom of the entire judicial system is considered "legal capital" (Landes and Posner 1976, as cited in Caldiera 1985). This conceptualization argues that recent decisions have more "value." It also assumes that later courts will make efficient use of the work of earlier courts when faced with the same question. This implies that courts wrestling with a gatekeeping decision should give more weight to the dominant and most recent trend among their peers.

Generally, then, the research question anticipates a pattern of convergence among courts, and a pattern where a decision is more likely to mirror the direction of the most recent jurisprudence on the particular question at hand. Furthermore, convergence should especially arise in the direction of acceptance. This is a "momentum theory" where the development of the science is interacting with the development of the law. The more the law keeps the science around, the more the science will develop and grow in popularity and validity. Thus, much like the horse race of front runners in elections, some acceptance could also begin to affect perceptions of validity, thus producing widespread general acceptance as a result of judicial validation. This is a theory of diffusion. A lack of convergence may signal the presence of an interaction pattern where law, attitudes, or other institutional/organizational variables are interfering with convergence.

As for the role of superior court decisions in judicial decision making, scholars have found that policy outcomes of lower courts are likely to favor the same parties or ideology as the decisions of the U.S. Supreme Court (Kilwein and Brisbin 1997). Caldiera (1985) found that much of the influence appears to be vertical rather than horizontal (see generally Canon 1973; Gruhl 1980, 1981b; Romans 1974). The influence of superior court decisions may vary depending on the degree to which courts feel accountable to those institutions either as an institution or as individuals. This is argued as part of the neo-institutional perspective (Brace and Hall 1989, 1990, 1992, 1993, 1995; Hall 1987, 1992, 1995) where the institutions of judicial selection and retention, and the institutions of oversight, can create an incentive for emulation.

Peer court decisions will necessarily affect the perceptions of judicial decision makers. Thus, we would expect to find a pattern of convergence. Similarly, in recognition of the role of recent decisions as influential, the direction of the most recent decisions will be weighted more heavily. In general, we would expect that peer court decisions should converge in a similar direction. In other words, judicial decisions with regard to DNA, polygraph, and syndrome evidence, while initially divergent, should become more alike as time passes. Thus, prior peer court decisions accepting the science can provide institutional political leverage for science in a current case. The same goes for prior rejections.

When it comes to judicial outcomes, institutional political leverage shares its power with organizational political leverage. Notwithstanding the importance of certain institutional positions in the judicial system, the politics of scientific evidence may also be a source of political leverage. This next group of hypotheses looks at the role of external organizations in political outcomes, and it anticipates that judicial outcomes will favor organized interests.

Professionalization

First, political leverage may arise from professionalization of the science at hand (certification, licensing, interpretive standards, level of autonomy from the legal community, etc.). If the expert witnesses are organized and certified, this assists in the framing and evaluation of their testimony. The perception of the validity and reliability of a particular science will increase with professionalization. Professionalization also aids science in meeting legal criteria by increasing arguments and evidence in favor of reliability, validity, and general acceptance. This is true because professional organizations develop standards for applying and interpreting scientific results and because they fund and publish relevant empirical information. Thus, when comparing cases, we may hypothesize that those cases with professionally certified witnesses would see more acceptance. This implies that for a particular science, professionalization of witnesses should significantly increase acceptance. Also, when comparing science to science, those with professional organizations will be more likely to gain the favor of the courts.

Complicating the politics of professional witnesses is the policy problem surrounding the institutionalization of forensic witnesses. Thompson (1997) conceptualizes forensic scientists as providing a service to a client. Primary clients are law enforcement agencies. He suggests that these incentives cause forensic scientists to act more like members of a trade guild—a guild that tends to be insulated from external scrutiny. He even goes so far as to argue that forensic scientists may be co-opted and adopt the goals of their clients as their own.

This can create a problem for courts. Courts require a neutral perspective on scientific methods, and a situation where scientists have an incentive to seek judicial approval of their methods is less than optimal. Thompson's observations set up an interesting test for professionalization where it is important to code for what "kind" of witness—internal or external to the legal system—is offering advice. There is a need to include this possibility of trade guild affiliation in order to distinguish between forensic scientists with a primary client base and academic scientists with more neutral goals. This also sets up a potential conflict between hypotheses. Whom will judges trust? Repeat-player status suggests that forensic scientists may enjoy extra deference because they know the game. Judges, however, may respect the opinions of scientists who are perceived as being outside the system. To test this, we would need a way to include some characteristic of the expert witnesses for each side to see if there is a correlation between experts and outcomes. There might also be a case where we can compare science to science based on the types of experts employed.

Another characteristic interacting with professionalization has to do with the nature of the science itself. Some science may present more ambiguity in results, and thus have more room for subjective interpretations. For example, fingerprint, DNA, polygraph, and spectrograph data are all subjectively interpreted. Syndromes, on the other hand, may have objective indicators a witness simply parrots to the court. For example, a label of battered woman syndrome is automatically attached to a situation that possesses the necessary indicators. It is counterintuitive that the hard sciences could suffer more than the social sciences simply because of interpretation, yet this could occur. Professionalization plays a significant role in terms of developing standards for interpretation. How should we expect professionalization to affect gatekeeping outcomes? First, the professionalization of witnesses and subsequent development of scientific and professional standards should increase acceptance of a particular science. Second, the presence of experts and professional expertise should increase acceptance. Third, the location of the profession relative to the criminal justice system (internal or external) may be correlated with judicial acceptance.

Third-Party Approval

Organizational political leverage may arise from approval by influential third parties. For example, in 1992 the National Research Council published its assessment of the science behind forensic DNA analysis. This synthesis of the state of the art should produce a change in judicial behavior, presumably in the direction favored by the contents of the report. Once reports surface, they should tip the balance in favor of the direction of the reports' recommendations. George and Epstein (1992) argue that judges cannot deviate too much from the larger political environment because they depend on other institutions for enforcement. In terms of third-party validation, this implies that the legitimacy of the agency may act on a court's own perceptions of a science as well as its perceptions of the way in which this validation will interact with public expectations. Third-party validation provides legitimacy to a decision favoring the same direction as the report. Thus, we would expect courts to decide in a manner congruent with such recommendations. This implies a pattern of conformity arising after the presence of third-party validation. A favorable report should produce favorable decisions. A mixed report should produce mixed decisions. In other words, favorable third-party validation should increase acceptance. Likewise, third-party reports should increase uniform decisions in the direction of the report.

Policy Advocates

Organizational political leverage may arise from the attributes of policy advocates who are stakeholders in the gatekeeping decision. Policy advocates refer to specific categories of litigants such as interest groups or corporations or amici individuals or groups. Policy advocates provide political information and cues, provide an impetus for necessary infrastructure for judicial use of a science, and have the skills and resources to compete successfully (Epstein 1994).

According to Kilwien and Brisbin (1997), American legal practice assumes that judges will respond to legal arguments drawn to their attention by litigants. Policy emulation should therefore be positively associated with the appearance of advocates for the policy doctrine. These perform three functions (Kilwein and Brisbin 1997): (1) they may force a state jurisdiction to make policy by sponsoring or otherwise supporting a case, (2) they may communicate the findings of other jurisdictions in their briefs, or (3) they may threaten appeal because of their resources (money, time, legal talent). Given these reasons, groups can play a role in judicial development.

When it comes to gatekeeping functions, policy advocates affect the framing and evaluation process. They provide information by virtue of their location within or outside of the legal community (insiders vs. outsiders). Policy advocates who are outsiders provide broader political support and legitimacy; insiders provide professional legitimacy and the expertise to successfully "lobby" the legal system. Policy advocates who are organized also provide an impetus for necessary infrastructure (labs, professional standards). One legal scholar provides compelling qualitative evidence that forensic labs, by virtue of the fact that they are stakeholders, have an incentive to move the judiciary in a direction favoring their interests (R. Epstein 2002). Policy advocates also provide information and political cues, and they have the resources and skills to compete successfully in political settings, setting up success in the courtroom (L. Epstein 1994).

A similar set of expectations recognizes the political power of broad coalitions of interests. The scope of the application of a particular science may affect the structure of policy advocacy. A science with broad applications may result in broad political coalitions supporting (or opposing) the science. For instance, the use of DNA for paternity suits as well as criminal investigations may improve its chances of political success. This expectation hypothesizes that broader science will have more organizational/institutional support and thus be able to provide the court with more persuasive cues.

The scope of application could also affect judicial outcomes simply because of the size of the policy domain. I hypothesize that a narrow application will feature fewer perturbations, uniform fact situations, and therefore uniform judicial treatment. When comparing one science to another, a change in function could lead to a change in judicial outcomes. For example, DNA for prosecution versus DNA for the defense (e.g., post-conviction DNA) has already rearranged the political framing and evaluation of DNA evidence in recent politics.

Policy advocates can affect gatekeeping decisions in a variety of ways. First, the more organized the policy advocates, the more likely the court will accept the evidence. This implies that because prosecutors are more organized than defendants, science supported by the prosecution should be accepted more often than science supported by the defense. This also implies that science where some elites are supportive and other elites are not supportive will experience more divergence than science where elites cooperate. Second, as the scope of the application increases, acceptance may increase if it widens the scope of political advocates/constituents. Acceptance may also decrease if a wider scope has the additional problem of increasing the uncertainty of the decision makers by sending them competing signals about the political leverage of the science.

State supreme court gatekeeping decisions require courts to frame and evaluate scientific evidence in a political context. This book examines an array of contributing factors to explore dominant patterns in this framing and evaluating process. Does doctrine impose a framework on judicial representations of reality? Do attitudes consistently frame and evaluate criteria in a predictable manner? Do systemic variables provide consistent leverage in the decision-making process? (E.g., Is there a pattern correlated with these variables?) Most fascinating for political science is that these variables are politically constructed. Furthermore, judicial outcomes clearly allocate values and resources, thus making them political decisions.

Gatekeeping Is a Judicial Outcome

Gatekeeping is a judicial outcome. Like most judicial decisions, the gatekeeping outcome is dichotomous; a jurisdiction either admits or rejects the evidence.

Assuming that judges and courts engage in cognitive processing as conceptualized, how could we model the way we expect the variables to affect the framing and evaluation of gatekeeping decisions? Explaining this outcome is most easily understood as searching for the factors responsible for a judicial decision.

As articulated above, expectations about judicial outcomes generally focus on the way three sets of factors (law, attitudes, and institutions/organizations) are related to court behavior. Law refers to the legal standards or tests used to evaluate the evidence. These standards may be difficult or easy, and they represent a low or high threshold for acceptability. If the legal threshold is low, it will presumably be easier for a science to clear this legal hurdle. If the legal threshold is high, it will presumably be difficult. Attitudes refer to the way preferences and beliefs of judicial actors can influence or direct a decision. If a judge's attitude toward the science is favorable, this can boost the likelihood of a science to be judicially accepted. Similarly, an unfavorable attitude from the judiciary could conceivably hinder admissibility. Institutions and organizations refer to those external actors that provide added leverage in the decision context (for instance, a third-party report supporting the use of a novel scientific technique in the courtroom). These factors work together to create a judicial gatekeeping outcome.

A Model of Judicial Gatekeeping Decisions

To illustrate these relationships, two decision environments are presented in figures 1.1–1.2. Each figure represents a different configuration of the factors discussed above. First, the legal threshold for admissibility is presented as a dotted line setting the bar for admissibility. Admissibility occurs when a case passes above the line. If a case remains below the line, it experiences denial. The threshold for admissibility (the legal standard at work in the jurisdiction) can be either a low threshold or a high threshold.

Figure 1.1 considers the situation where a case finds itself in a court with high attitudinal support, such as a case where the court is conservative and law enforcement is arguing for admissibility. The threshold for admissibility for the jurisdiction may be high or low, depending on the legal standard used in the state. If the threshold is low, the case will require only a small amount of political leverage to clear the hurdle because the court is already predisposed to support admissibility. In this way, high attitudinal support has moved this case close

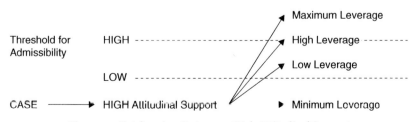

Figure 2.1 Gatekeeping Outcomes: High Attitudinal Support

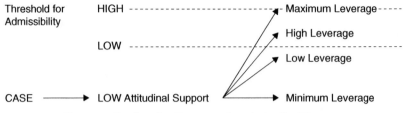

Figure 2.2 Gatekeeping Outcomes: Low Attitudinal Support

to a low threshold, thus requiring only a small amount of political leverage to clear the legal hurdle. If the threshold is high, the case will require more political leverage to clear the hurdle.

Figure 1.2 changes the location of the case by changing the attitudinal support of the court. In Figure 1.2 the case finds itself in a court with low attitudinal support, such as a case where the court is conservative and the defendant is arguing for admissibility. This has the effect of moving the case further from the threshold, thus requiring *more* political leverage to clear the legal hurdle. For cases in this situation, significant political leverage is required for even the lowest legal threshold. If a jurisdiction has a high threshold for admissibility, the case can only clear it with the highest amount of institutional and organizational leverage because it has so little attitudinal support from the court. In other words, when faced with low attitudinal support it would take a great deal of help, such as a skilled attorney, plenty of expert evidence, and perhaps a favorable third-party report, to be admitted in a jurisdiction with a high threshold. These figures demonstrate the way law (as a threshold for admissibility), attitudes, and political leverage can work together to produce differential judicial outcomes.

Several assumptions are at work in this model. While somewhat technical, it is helpful to explicitly articulate these assumptions to make relationships and implications clear. Judicial gatekeeping decisions have only two possible outcomes, which can be conceptualized as clearance or denial. *Admittance* is the judicial acceptance of a way of knowing (science) for legal purposes. This term was selected because it implies both approval and the concept of clearing a bar or threshold. *Denial* is the judicial rejection of a way of knowing (science) for legal purposes. It was selected because it implies disapproval. The attainment of clearance or denial is dependent on three sets of factors: law, attitudes, and institutional/organizational leverage.

Several assumptions underlie the interaction of these factors with judicial outcomes. First, neither law nor attitudes is sufficient for denial or admittance. Law alone cannot determine clearance or denial. Law here refers to the use of a legal standard or threshold to determine admissibility. The existence of a legal standard implies at least some hurdle to be overcome for admittance to occur. The existence of a legal standard as a decision-making tool also implies that some science must meet or exceed the standard. For these reasons, it can be assumed that no legal standard is so low as to be non-existent. It can also be assumed that no legal standard is so high as to preclude any admittance. Likewise, judicial attitudes alone cannot determine admittance or denial. Attitudes represent the attitudinal support a court may have for clearance. While important in the framing and evaluation of case material, attitudes alone cannot result in admittance or denial. No matter how strong the preference to admit or deny the evidence, a court must have an item of institutional support (some kind of substantiation) for the decision. For this reason, attitudinal support, no matter how great it is, will still require additional support from institutional/organizational leverage (expert witness testimony, third-party reports, etc.) to pass a legal threshold, even a very low legal threshold.

Second, institutional/organizational leverage can be sufficient for admittance or denial, irrespective of law or attitudes. Institutional/organizational leverage, if high enough, can overcome even the highest legal threshold or lowest attitudinal support. Likewise, if institutional/ organizational leverage is low enough, denial will occur, regardless of a low legal threshold or high attitudinal support. In other words, some institutional/organizational support is necessary for admittance to occur.

Third, the magnitude of the effect of law or institutional/organizational leverage is at least as much as the magnitude of the effect of attitudes on a judicial outcome. In other words, attitudinal support, while part of the story, would not be nearly as important as the location of the legal threshold or the amount of institutional/organizational support.

Fourth, the legal threshold is fixed prior to the case at hand. Jurisdictions have one legal standard of admissibility for novel scientific evidence as the law of their land. This legal threshold is not changed to accommodate the case at hand. On occasion, courts do use a case as an opportunity to reexamine and perhaps even change their legal standard. It can be argued, however, that (1) this is very rare, (2) changing a legal standard often has little to do with the case at hand. Rather, the case has presented an opportunity to recalibrate their doctrine, and (3) changing the legal doctrine recalibrates the threshold for all future cases as well, so it cannot be considered case specific or even case dependent (though it is acknowledged that courts may see a case as particularly suited for a new doctrinal articulation).

Fifth, attitudes can have an impact on the effect of institutional/organizational leverage. The most important role for attitudinal support is its role in the framing and evaluation of institutional/organizational support. For this reason, cases could theoretically have identical amounts of institutional/organizational leverage and an identical threshold, and still have different outcomes. This is because of the interaction of attitudes with institutional and organizational leverage.

These assumptions have several implications. First, each variable may account for the variance independently of the other variables. As stated above, two cases with the same legal threshold and institutional/organizational support could have different outcomes because of a variance in attitudinal support. Likewise, two cases with the same attitudinal support and the same legal threshold could have different outcomes because of a variance in institutional/organizational leverage.

And two cases with the same attitudinal support and institutional leverage could have different outcomes because of a variance in the legal threshold. Second, it implies that the highest amounts of institutional/organizational leverage will be sufficient for *admittance* and that neither the lowest amounts of attitudinal support nor the highest legal threshold is sufficient for *denial.*

Gatekeeping and the Metaphorical Context

We return to our metaphor of the gatekeepers along the ancient city walls. If instilled with meaningful power, these gatekeepers would likely operate under the same models posited above. First, they would operate under the legal standards of the jurisdiction—a liberal standard would presumably allow more strangers into the city than a conservative standard. Second, they would be unable to escape their attitudes toward particular types of strangers and the politics those strangers represent—this favorable or unfavorable attitude toward the perceived benefit or harm of allowing the stranger to enter would surely be a detectable pattern. Third, gatekeepers would be more likely to admit strangers who enjoyed institutional or organizational advantages, for example those in the company of known individuals (repeat players) or those who had been deemed beneficial by an official report (third-party validation). Each of these expectations recognizes that the city gatekeepers are operating under limited cognitive capacity and ambiguity, in a manner similar to judicial gatekeepers. Certain predictable outcomes are possible when the expectations of these political models are specified.

Let us return once more to judicial processing of scientific information. To an observer, each court is judging the same *way of knowing*; thus differential outcomes must arise from something other than the science in question: the application of different standards/criteria, the pursuit of different goals/policy outcomes, and/or the use of different perceptions created by institutional/organizational variables. Each of these is politically defined in the U.S. judicial system. In political science, these three sets of variables represent three theoretical approaches to judicial outcomes: the legal model, the attitudinal model, and institutional/organizational models. These explanations are the likely starting place for solving this empirical puzzle.

The analysis of these models in the gatekeeping context is important for theoretical development. Gatekeeping is policy making at the jurisdictional level. Unlike civil liberties rule making, however, gatekeeping decisions carry less political information. Policy makers must perform under conditions of political and scientific uncertainty. The external forces (as opposed to internal attitudes) creating the "cues" used by decision makers are more powerful in this context, and the role of these forces in judicial outcomes ought to be stark.

This analysis is also important for public-policy development. The theoretical concerns matter because of the *political* implications. Is the system functioning correctly? Does it do a good job testing the viability of certain "ways of knowing"?

Judicial processing of scientific information will depend, at least in part, on the political process. The politically constructed variables creating the judicial outcomes are a product of a political process rather than a scientific process. The question becomes, How does that process develop? What are the mechanisms at work?

At the center of the judicial system are human decision makers. Explanation comes from describing that which causes their decisions. This is both individually conceptualized (attitudes) and systemically conceptualized (law, institutions, scientific applications, etc.). Thus, while the internal forces operating on individuals (attitudes) are important, the external forces operating on those individuals (law, institutions, organizations), as well as the political processes producing those external forces (namely politics—for the institutions and organizations are politically constructed), are equally important. Lawyers use the legal model to anticipate judicial outcomes, and law review research demonstrates a pattern of doctrinal variance correlated with judicial outcomes. Political scientists similarly use the attitudinal model to anticipate judicial outcomes, and the judicial branch literature has demonstrated correlations between attitudes and judicial outcomes. Furthermore, institutional and political variables will affect the judicial processing of scientific information. In the judicial decision-making context, sources of leverage may include: repeat-player status, appellant status, lower court decisions, peer court context, superior court context, witness professionalization, and the organizational structure of policy advocates. Each source of leverage can vary in terms of magnitude and direction. These variables may be empirically associated with judicial outcomes because they add to the perception that science meets doctrinal or ideological criteria.

The explanation of judicial outcomes is a complex undertaking. By conceptualizing the problem in terms of the factors affecting cognitive processing, the following analysis demonstrates important relationships among variables, framing, and evaluation. The primary goal is to find which patterns (legal, attitudinal, institutional, organizational) are present in the judicial processing of scientific information. Theoretically and politically these findings should explain many of the forces behind the empirical situations. It should also explain the way in which judicial treatment of science is due to political rather than scientific factors.

CHAPTER 3

FORENSIC DNA

LAW ENFORCEMENT IN THE LABORATORY

As a very important *way of knowing*, forensic DNA evidence presents the first gatekeeping mystery for exploration. The DNA controversy officially began in 1989 with the first American appellate case. The first high court to render a decision on the admissibility of DNA evidence was the Supreme Court of Virginia. In *Spencer v. Commonwealth*, 384 SE 2d. 785 (1989), the defendant, Mr. Spencer, appealed a capital murder and rape conviction on the grounds that DNA evidence should not have been admitted at trial "because the commonwealth failed to establish its reliability and its general acceptance in the scientific community" (797). The Supreme Court of Virginia undertook an extensive review of the expert testimony at trial and agreed with the lower court that "DNA testing is a reliable scientific technique and that the tests performed here were properly conducted" (797). That same year, just two months later, the Supreme Court of Minnesota in *State v. Schwartz*, 447 NW 2d. 422 (1989) rejected DNA evidence after an equally extensive review of the science and the expert testimony.

As other states began to hear DNA cases, they tended to side with either Virginia or Minnesota. Eventually, all the states accepted DNA evidence, but only after a protracted, jurisdiction-by-jurisdiction battle. From 1989 to 2003, state supreme courts heard 153 cases on the scientific validity and reliability of DNA evidence (see appendix A). Twenty-seven decisions found DNA evidence (generally) to be invalid or unreliable (or unproven as valid and reliable). This is a gatekeeping mystery if there ever was one. Why did some courts reject while others accepted? Why did it take so long? How could the courts reverse course once they had ruled a certain way?

As noted in chapter 2, political scientists have developed several explanations for judicial activity, including the law of the jurisdiction, the attitude of the court, and the various institutional and organizational advantages of the parties in the case. The pattern of judicial outcomes and the related factors for DNA evidence provide an important starting point for gatekeeping analysis.

A Science Lesson for the Gatekeepers

Forensic DNA evidence is a complicated matter for judicial gatekeepers to consider. With regard to DNA evidence, there are actually three *ways of knowing* (science) involved: the theory, the technique, and the interpretative statistics employed. State supreme courts faced with the task of deciding whether forensic DNA is an admissible form of science must rule on each of these ways of knowing. The three ways of knowing mirror the general steps in DNA testing: "(1) Creating a DNA print or profile of a sample, (2) Determining whether the prints or profiles of different samples match and (3) If samples match, computing the probability of a random match" (*State v. Bible*, 858 P.2d 1152, 1180 [Ariz. 1993]). I will describe each part briefly to provide background and technical understanding.

Theory

The DNA molecule is a double-stranded molecule composed of nucleotide bases that are paired together. The sequence of these bases determines the message the DNA carries. The theory of DNA is simply that no two people share the same exact DNA sequence (except for identical twins). These sequences of DNA, found in every cell of the body, are unique to each individual. In a criminal situation, crimes involving forensic evidence (murder, rape, larceny, etc.) may utilize DNA evidence. The most common tests involve matching a suspect with the biological evidence (blood, semen, hair) found at the crime scene. In theory, the DNA found at the scene should only match the individual from whom it originated. Labs, however, cannot test all three billion base pairs on a complete DNA molecule. Rather, labs test certain locations (loci) known to vary among individuals. These regions are often referred to as polymorphic regions on the DNA molecule. A typical DNA test will look for matches in several polymorphic regions (usually seven to thirteen loci). If the DNA from the suspect contains the same polymorphisms at the same loci as the DNA from the crime scene, a "match" will be declared. In this sense, the "theory" of DNA matching hinges on the assumption of variance across individuals, and thus, DNA analysis provides a fair amount of certainty of identity. For this reason, DNA evidence is often compared to fingerprint evidence. Both forensic tools involve theoretical assumptions of uniqueness and reliable measurement of that uniqueness. Fingerprint analysis paved the cognitive and legal path for the theory of DNA analysis, and DNA evidence is often referred to as DNA fingerprints.

Technique

Once the theory of DNA is accepted, courts undertake the evaluation of the techniques involved in DNA analysis. These judgments require decision makers to find laboratory techniques valid and reliable for legal identification purposes. From 1989 to 2003, courts evaluated two types of DNA analysis: restriction-fragment length polymorphism (RFLP) analysis and polymerase chain reaction (PCR) analysis.

RFLP analysis involves several steps. First, restriction enzymes are applied to the sample. These enzymes cut the DNA into fragments at the location of particular sequences, called recognition sites. These recognition sites vary among individuals. If a site does not exist, the strand of DNA will be uncut, resulting in a longer strand for that individual. These variations of DNA fragments are called restriction-fragment length polymorphisms. Second, the size of the variation is visualized by placing the fragments in an agarose gel and applying an electric current. The fragments move with the current across the gel at various speeds depending on their size (smaller fragments go farther). This electrophoresis process separates the DNA by size to create bands of DNA. Third, to "see" these bands, radioactive probes are introduced. They bind to the fragments, and photographic film is overlaid to expose the locations of the fragments. DNA analysis looks at banding patterns to compare the RFLPs of two separate samples (e.g., one from the suspect and one from the crime scene). If the locations of the fragments are the same, a "match" may be declared.

One major drawback of RFLP analysis is that it requires a relatively large, un-degraded DNA sample. Many crime scenes offer only tiny amounts of DNA material for analysis. PCR (polymerase chain reaction) amplification provides a method for increasing the size of the sample. The PCR method was described as follows by the National Research Council in 1992:

> PCR allows for a million or more copies of a short region of DNA to be easily made. For DNA typing, one amplifies (copies) a genetically informative sequence, usually 100–2000 nucleotides long, and detects the genotype of the amplified product. Because many copies are made, DNA typing can rely on methods of detection that do not use radioactive substances. Furthermore, the PCR amplification technique permits the use of very small samples of tissue or bodily fluids—theoretically even a single, nucleated cell. (5–6)

Once the sample is created, methods of match analysis are similar to RFLP techniques and may even be identical.

Statistics

Once the theory of DNA and the laboratory techniques have been accepted by the courts, there is still a question about the reliability and the validity of the probability statistics used to decide the uniqueness of the match. These statistics are used to describe the likelihood that the DNA came from a third person, even though it matches. In other words, these statistics provide the probability that it is a random match. If the probability of a random match is relatively low, the likelihood the sample came from the suspect increases. These statistics first look at the frequency of certain polymorphisms (alleles) in the target population, usually determined by the race of the suspect. These frequencies are then multiplied by the frequency of the genotype using the product rule. This provides the probability that a particular polymorphism occurs in the target population.

The last step is to look at the probability of all polymorphisms (one from each loci) occurring in the same individual. This is done by using the product rule to multiply the frequencies of each polymorphism with each other. These calculations generally result in very small chances of a random match, and they are usually articulated in court as "one in several million" or higher.

The creation of the target population frequencies and the use of the "product rule" has been a source of judicial (and scientific) controversy in many cases involving DNA analysis. The target samples are often contested as unrepresentative or untested. The product rule has been challenged on the grounds of probability assumptions of independence.

The Challenge of DNA

As the above discussion illuminates, the admissibility of DNA evidence involves essentially three judgments about science. Courts are required first of all to declare the validity and reliability of the theory. Next, they have to look at the reliability and validity of complex and highly technical methodologies—specifically RFLP analysis and PCR amplification. Finally, courts have to decide if the statistics used to give meaning to match evidence are reliable and valid measurements of occurrences within the population. The legal literature on the "controversy" of DNA evidence does not always recognize the complexity of this three-tiered analysis. Indeed, the variance among jurisdictions is often a result of rulings on different levels: theory, technique, or statistics.

Even so, some political questions remain important for the following analysis: when courts do vary in their acceptance of DNA evidence, what factors are most important for determining outcomes? On the surface, one might expect the question to depend on the "quality" of the science. The "quality" of science, however, is actually a result of three political processes.

One process is the positioning of supporting apparatus and players in a way conducive to a successful legal fight. This is the political process responsible for the proliferation of labs, professionalism, law enforcement practices, research, and reports from the scientific community, all of which contribute to the view of that science as valid and reliable. Political allocation of values and resources clearly determines the way the "science" will negotiate the legal system by determining the availability of labs, experts, and knowledge. Indeed, the story of DNA jurisprudence must recognize the concerted effort of the federal government through the FBI and the Department of Justice to provide the necessary infrastructure for DNA testing and testimony. According to a report by the George W. Bush White House in 2003, the federal government is solely responsible for the successful development of DNA technology and infrastructure for law enforcement purposes (see generally *Advancing Justice Through DNA Technology 2003*, a policy book posted on the Internet by the Department of Justice with the seal of the President of the United States).

A second process is the process whereby standards for admitting scientific evidence are developed and interpreted. The proliferation of the *Frye* standard, the *Daubert* standard, and the relevancy standard (FRE 702) all determine the threshold acceptability science must overcome to become admissible. These standards are developed in a political manner by political processes in the judicial system. The political process also provides for variance from jurisdiction to jurisdiction.

A third process is the manner in which individual and collective courts receive and evaluate information for decision making. The cognitive process involved in human decision making leaves plenty of room for politics—either in the pursuit of conscious policy goals or in the unconscious sorting of information according to ideological schema. It might be expected that judicial outcomes could have more to do with human factors on the court than with the law or the science.

With this introduction to DNA analysis, it is apparent that the courts of the 1990s had their work cut out for them. They were trying to determine what is reliable and valid.

The Gatekeepers and DNA Evidence
General Findings

From 1989 to 2003, there were 153 cases where state supreme courts ruled on the scientific reliability or validity of DNA evidence. These cases were collected using a LexisNexis search of all high court cases containing the phrase "DNA." Each case was then briefly read to isolate wherein the court ruled on the validity/reliability of the science (theory, technique, or statistics) involved in DNA analysis. As state supreme court cases, these DNA decisions are decisions where the state supreme court engaged in policy making for the jurisdiction. Their decisions resulted in new gatekeeping policy for all trials statewide. These cases are not a sample; they compose the complete picture of analysis.

Once collected, each case was coded for the holding with regard to admissibility of the DNA evidence in question and particular holdings with regard to theory, technique, and statistics. In many cases, the theory or the technique would pass muster while the statistical calculations fell short. The following descriptive statistics break this down by year, state, and holding.

In table 3.1, the initial pattern of acceptance and rejection is shown across time. Table 3.1 is a timeline, with state supreme court decisions represented by the state abbreviation. As noted above, in 1989, Virginia had a decision accepting DNA evidence while Minnesota rendered a decision rejecting DNA evidence. By examining this timeline, several observations are immediately notable. First, the majority of decisions (126) have accepted DNA. Of those states rejecting DNA evidence, all the decisions before 1995 represent jurisdictions that later admitted DNA evidence. The decisions after 1995 represent jurisdictions with a mixed DNA track record. Some had previously admitted DNA evidence, while others will continue to wrestle with admissibility issues depending on the case. Second,

TABLE 3.1

ADMISSIBILITY OF DNA IN STATE SUPREME COURTS
(DECISIONS BASED ON SCIENTIFIC RELIABILITY OR VALIDITY)

Inadmissible		Admissible
MN	1989	VA, VA, WV
ID	1990	GA, NC, SC, VA
AL, MA, MN	1991	AR, CA, IN, IN, IA, KS, MO, SD
CT, MA, MS, NE, NH	1992	AR, AR, FL, HI, IN, MN, MS, OH, TX, TX, WY
AZ, DE, MA, WA	1993	CO, IN, KS, MN, WY
CT, CT, NE	1994	CT, MA, MN, MT, NH, NM, NM, NY, VT, WA
FL, VT	1995	CO, FL, GA, ID, KS, KS, KY, MT, OK, SC, SD, TX, WA
AL, LA, MD	1996	AZ, AR, FL, IL, IL, MD, MS, MO, OR, RI, SC, SD, TX, WA, WI, WA, WA
FL	1997	AZ, FL, IL, IN, KS, KY, ME, MA, MA, MA, NE, NE, NJ, SD, TN, WA
AL, CA	1998	DE, IN, IA, IN, KY, MS, NE, NV, NM, OK, OK, PA
	1999	KS, KY, LA, MS, MS, SC
	2000	GA, LA, MS, ND, TN, TX
	2001	CO, CO, CT, LA, MA, SC, UT, WA
FL, MN	2002	AR, MD
	2003	AL, CA, NH, MN, MN
Total 27 cases (18%)		126 cases (82%)

the amount of state activity over time is not steady. The timeline shows that DNA high court decisions first appeared in 1989 with four jurisdictions considering the question. The years 1992–1998 were "active" with 103 decisions forthcoming, and five of these years averaged over fifteen cases per year. After 1998, the jurisprudence slows down noticeably, with only twenty-eight cases occurring from 1999 to 2003.

The timeline also shows that states vary in the number of DNA decisions undertaken during this period. Two states have had no high court decisions on the validity and reliability of DNA evidence in criminal trials. Twelve states register one DNA decision. The average number of decisions is approximately three decisions per jurisdiction, and indeed the mode number of decisions is three (thirteen states). A few states revisited questions of reliability or validity on numerous occasions; Florida, Indiana, Massachusetts, Minnesota, Mississippi, and Washington each had seven or eight DNA decisions from 1989 to 2003.

These general findings create the puzzles investigated later in the analysis: Why did some states admit while other states rejected? Why were some states hemming and hawing more than others? What is significant about the noticeable reduction in rejections after 1996? Before we can begin the quest for causation, it is necessary to take a closer look at the nature of the holdings with regard to DNA theory, techniques, and statistics.

Theory

The theory of DNA has met with little controversy at the jurisdictional level. No court decision has held the theory of DNA to be invalid or unreliable. In fact, even decisions where techniques or statistics are called into question are careful to state that the theory of DNA is generally accepted. This means a scientific theory can be approved by the court, even while its techniques are rejected.

A handful of dissenters on various courts did feel compelled to state their reluctance to wholeheartedly embrace DNA theory. None was more compelling than the succinct opinion of Justice Dore (joined by Justice Utter) on the Washington Supreme Court: "I dissent. DNA testing is not reliable; it does not pass the *Frye* standard, and it is not admissible" (*State v. Cathon*, 846 P.2d 502 [Wash. 1993]). On the whole, most cases say something similar to the court's opinion in *Ex Parte Perry*, 586 So.2d 242, 244 (Ala. 1991): "The general scientific theory underlying DNA print analysis is almost universally accepted in the scientific community." Other decisions reiterate the lack of challenges to the theory in a manner similar to the following: "Defendant does not challenge DNA testing in toto. Indeed, defendant concedes general acceptance of the underlying theory of DNA testing and its research and diagnostic purposes" (*State v. Bible*, 858 P.2d 1152, 1179 [Ariz. 1993]). Thus, little legal, scientific, or political controversy surrounds the theoretical proposition and assumptions of DNA typing.

Technique

DNA fingerprinting techniques experienced more controversy. Judicial outcomes with regard to forensic DNA techniques revealed jurisdictions that are hesitant to admit evidence. Decisions on the science of DNA evidence generally, restriction-length fragment analysis (RFLP), polymerase chain reaction (PCR) evidence, and band shifting are included in the conceptualization of DNA techniques. Overall, twelve cases ruled DNA techniques inadmissible for lack of validity or reliability while ninety-one cases ruled the techniques acceptable. Of those cases, four decisions simply ruled on "DNA" or "DNA techniques" (Del. 1993; Fla. 2002; Idaho 1990; Miss. 1992). Two cases ruled against RFLP analysis (Conn. 1992; Neb. 1992).

Two cases ruled against band-shifting analysis (Fla. 1995; La. 1996). Four cases ruled against PCR analysis (Fla. 1997; Md. 1996; Minn. 2002; Neb. 1994). These results stand in contrast to fifty-eight rulings specifically recognizing the reliability and validity of RFLP analysis and the forty decisions favoring PCR. It is also interesting to note four jurisdictions that originally ruled against RFLP or PCR and then subsequently admitted the technique (Conn. 1997; Md. 1996; Minn. 2003; Neb. 1994). It is notable that roughly half of the state supreme court decisions rejecting DNA evidence did so on the basis of the technique employed.

Statistics

Judicial outcomes with regard to forensic DNA probability methods and statistics are by far the most controversial decisions. Seventeen decisions rejected the reliability and validity of statistics. Fourteen of those decisions accepted DNA theory and DNA techniques only to reject the probability statistics. Fourteen of them also occurred before 1996. The use of statistics in DNA testimony is controversial because methods for calculating probability present extra dimensions for judicial dissatisfaction. These are not laboratory questions. These are questions about the way match statistics are generated and presented to juries. For example, the application of statistics can be challenged in terms of the sample used to create the frequencies for the target population. Opponents can also challenge the assumptions of independence for each polymorphic location, rendering the use of the product rule invalid. Even with these potential weaknesses, favorable decisions about DNA probability statistics still outweighed unfavorable ones; sixty-nine cases upheld the reliability and validity of statistical methods.

Overall, judicial behavior indicates that theory is the least controversial "way of knowing" for the legal system. Specific scientific techniques are more controversial, especially the PCR technique, which was rejected in 10 percent of relevant cases. Statistical application is the most controversial part of DNA analysis: it was rejected 20 percent of the time.

BLACK ROBES IN ACTION

Besides jurisdictional variance with regard to particular rulings, interesting patterns of jurisdictional behavior also emerge from the descriptive statistics (see table 3.2). First, there is a distinct group of states whose initial ruling to admit DNA evidence was never revisited. These *"single-decision"* jurisdictions had only one decision on the reliability and admissibility of DNA evidence. Those decisions all favored admitting DNA evidence and they appeared to set a precedent as the sole judicial discussion of the science of DNA evidence for that jurisdiction. Twelve jurisdictions exhibit this phenomenon of initial, complete, and unchallenged acceptance: Hawaii, Maine, New Jersey, New York, North Carolina, North Dakota, Ohio, Oregon, Rhode Island, Utah, West Virginia, and Wisconsin. These single-decision cases occur across the timeline with early precedents (two before 1992) as

TABLE 3.2

DECISION PATTERNS FOR SINGLE-DECISION AND DYNAMIC JURISDICTIONS

Single-Decision to Admit DNA*

Year	State(s)
1989	West Virginia
1990	North Carolina
1992	Hawaii, Ohio
1994	New York
1996	Oregon, Rhode Island, Wisconsin
1997	Maine, New Jersey
2000	North Dakota
2001	Utah

*No jurisdiction had a single, unfavorable DNA decision.

Dynamic Jurisdictions and DNA

State	(Early Rejection Decisions) Subsequent Acceptance Decisions
Alabama	(1991, 1996, 1998) 2002
Arizona	(1993) 1996, 1997
California	(1998) 1999, 2003
Connecticut	(1992, 1994) 1994, 2001
Delaware	(1993) 1998
Idaho	(1990) 1995
Massachusetts	(1991, 1992, 1993) 1994, 1997, 1997, 1997, 2001
Minnesota	(1989, 1991) 1992, 1993, 1994; PCR: (2002) 2003, 2003
Mississippi	(1992) 1992, 1996, 1998, 1999, 1999, 2000
Nebraska	(1992, 1994) 1997, 1998
New Hamp.	(1992) 1994, 2001
Pennsylvania	(1994) 1998
Washington	(1993) 1994, 1995, 1996, 1996, 1997, 2001

well as latecomers (four after 1996). The presence of early decisions implies that some jurisdictions do not feel the need to revisit initial precedents with each new technique or controversy surrounding DNA. It also suggests this single-decision phenomenon is not limited to latecomers who simply wade into the calm waters of a receding controversy.

Second, there was a group of states where initial decisions and subsequent decisions differed. This group of *dynamic* jurisdictions is distinguished by the fact that all initial decisions against DNA evidence were eventually overruled in favor of DNA evidence. This significant change in precedent suggests that a legal or political threshold for acceptance was finally overcome in the jurisdiction. It may also be evidence of convergence. These dynamic jurisdictions present an interesting comparison to the "single-decision" districts with respect to timing. The initial rejections of these states are clustered in the early years of DNA jurisprudence. All but one rejection decision occurred before 1995. By contrast, the acceptance decisions of single-decision jurisdictions occur relatively evenly across the timeline, with half before 1996 and half after 1996. This may mean those who "got it right" were not forced to revisit the question, while those who "rejected" DNA were continually pressured to readjust their jurisprudence.

Third, there was a group of states with multiple, favorable decisions in the direction of acceptance. This group of *monotonic* jurisdictions is characterized by steady jurisprudence in favor of admissibility. Twenty states are monotonic with regard to DNA evidence: Arkansas, Colorado, Georgia, Illinois, Indiana, Iowa, Kansas, Kentucky, Maryland, Montana, Nevada, New Mexico, Oklahoma, South Carolina, South Dakota, Tennessee, Texas, Virginia, and Wyoming.

Taken together, the behavior of single-decision jurisdictions, dynamic jurisdictions, and monotonic jurisdictions suggests strong evidence for an overall pattern of convergence with regard to DNA evidence. Furthermore, the direction of this convergence is toward admissibility. One wonders how well this mirrors the jurisdictional patterns of other types of science. Is convergence the norm? When might prolonged divergence appear? The other question is why did some courts reject DNA evidence? Before we compare judicial outcomes for DNA to judicial outcomes for other types of science, we will look at the possible factors responsible for decisions and decision patterns.

Explaining DNA Gatekeeping: Legal Considerations

One clue to judicial behavior suggests that gatekeepers are operating according to some standard of who (and what) can (or cannot) be admitted. If the rules of admissibility vary from jurisdiction to jurisdiction, this would presumably affect the chances of admittance. In chapter 2 this was conceptualized as a "threshold" at work in a jurisdiction. Jurisdictions with lower thresholds (or lower legal standards) will presumably experience more acceptance than jurisdictions with higher thresholds.

By way of review, there are essentially three legal standards used in the judicial admittance of scientific evidence: the relevancy standard (state versions of the Federal Rules of Evidence Section 702, hereinafter FRE 702), the general acceptance standard (*Frye* standard), and the *Daubert* standard (four potential criteria for a judicial decision based on the *Daubert* case). The relevancy standard requires that scientific evidence be relevant to the case, presented by a qualified scientific expert, and provide more probative than prejudicial value. The general acceptance test requires that the court find the scientific theory or technique to be "generally accepted" by the relevant scientific community. The *Daubert* standard is broader, allowing the judge to consider general acceptance, as well as testability, peer-reviewed tests, and known rates of error.

These standards are generally viewed by legal scholars as part of a continuum of thresholds the science must meet to be admitted. FRE 702 has the least scientific rigor. The *Daubert* standard contains several elements of scientific rigor yet still provides the gatekeeping court with more discretion than the stringent "general acceptance" test. This is easy to illustrate. Under the *Daubert* standard a scientific method does not have to pass all four criteria. This means a new or novel technique that has not gained "general acceptance" in the scientific community might be admissible if it meets the other three criteria. The *Frye* standard is considered the most difficult standard. Courts can define "general acceptance" in a strict manner, effectively rejecting evidence if there is a hint of dissent in the scientific community.

A handful of states rejected all three of these standards. The court in Georgia consistently ruled in favor of expansive trial court discretion, asking only that the judge make a determination finding the science valid, reliable, and acceptable. The Georgia court explicitly rejected any standard resembling a "general acceptance" test: "The standard in Georgia is that we do not want to count heads in the scientific community" (*Caldwell v. State*, 393 S.E.2d 436, 441 [1990]). Virginia is another state with its own standard. Its jurisprudence simply requires a finding that the science is "proven reliable." Trial court judges are given broad discretion. Nevada courts did not engage in any type of legal analysis for DNA questions. They simply upheld the admittance of the evidence.

In terms of DNA evidence, the legal standard at work in a jurisdiction was expected to influence judicial outcomes. The more liberal the standard, the more likely the jurisdiction would find the science acceptable. The more difficult the standard, the more likely scientific evidence would be rejected. Even the strictest standard, the *Frye* "general acceptance," however, has an important built-in dynamic: time and diffusion of a novel scientific technique can result in growing acceptance by the scientific community. So, even in *Frye* districts, we may expect to see a pattern where *Frye* districts that initially rejected DNA later found it to be "generally acceptable" simply because of widespread use and acceptance by the scientific community over time. It is also possible that "liberal" standards provide science with a legal foothold that eventually allows it to overcome more conservative standards. For instance, as some jurisdictions admit DNA, the use and acceptance of the technique grows, possibly resulting in later admittance by more reluctant jurisdictions.

First, we can look at jurisdictional behavior. Jurisdictions have varied greatly in the legal standard they have applied to DNA evidence. Some jurisdictions maintained specific legal standards in their DNA jurisprudence. The general acceptance standard (*Frye* in most jurisdictions) is the most popular. Twenty-one jurisdictions maintain this standard. The relevancy standard (state variants of FRE 702) is maintained by nine jurisdictions. The *Daubert* standard was maintained in two states.

Other jurisdictions experienced significant shifts in their legal calculi. The most popular shift was moving from requiring a finding of "general acceptance" (*Frye*) to requiring a *Daubert* finding where general acceptance is but one path to admissibility. This was popular because the U.S. Supreme Court moved away from its 1923 *Frye* test by adopting the *Daubert* test in 1993. Thus, several jurisdictions decided to revisit their own requirements in light of the U.S. Supreme Court ruling. Nine formerly *Frye* jurisdictions switched to the *Daubert* standard at some point in their DNA jurisprudence. Two states decided to abandon *Frye* for the more discretionary FRE 702 standard. One state decided to move from its FRE 702 to a *Daubert* standard after 1993.

This variance within and among jurisdictions provides a ready context for testing a legal standard hypothesis. To what extent is the variance in judicial outcomes correlated with the legal standard employed by the jurisdiction? First, we can disaggregate cases from jurisdictional clusters and look at individual cases. Here, all decisions are treated as independent of jurisdiction (see table 3.3). This reveals that, indeed, the majority of "no" decisions are clustered in those decisions employing the "general acceptance" (*Frye*) test. Indeed, 80 percent (twenty-one of twenty-six) of the decisions rejecting DNA evidence used the *Frye* standard. It is also interesting that 23 percent of *Frye* decisions rejected DNA evidence while only 14 percent

TABLE 3.3
COMPARING LEGAL STANDARDS AND JUDICIAL OUTCOMES

Legal Standard in the Decision		Cases Admitted	Rejected
Frye or "general acceptance"	93	77% (72)	23% (21)
Daubert (one of four criteria)	22	86% (19)	14% (3)
FRE 702	29	93% (27)	7% (2)
Other	9	100% (9)	0% (0)

Likelihood Ratio 8.758 (p = .033)
Kendall's tau-b .198 (p = .002)

of *Daubert* decisions, 7 percent of FRE 702 decisions, and 0 percent of "other" decisions rejected DNA evidence. From this evidence, one can conclude that the legal threshold has at least some effect on judicial admissibility rates.

A second cut of the analysis is to look at those dynamic jurisdictions where the decisions shifted from rejection to acceptance (see table 3.4). In order to see if the changes were the result of a change in legal standard, table 3.4 presents those jurisdictions that changed DNA. The goal is to examine whether a shift in judicial acceptance correlated with a shift in the legal standard employed by the jurisdiction. Thirty of those admitting decisions under *Frye* are in jurisdictions where other *Frye* decisions (usually occurring prior) were rejecting decisions. This means that the use of the *Frye* standard appeared to allow for a movement toward acceptance, but it clearly slowed down acceptance compared to other standards. Of those districts with a dynamic admissibility pattern (where initial rejection was later changed to acceptance), almost all of them started with the *Frye* analysis. Three jurisdictions (Alabama, Massachusetts, and New Hampshire)

TABLE 3.4

JURISDICTION PATTERNS AND LEGAL STANDARDS FOR THOSE INITIALLY REJECTING DNA EVIDENCE (DYNAMIC JURISDICTIONS)

State	(Early Rejection) Subsequent Acceptance	Legal Standard Employed
Alabama	(1991, 1996, 1998) 2002	(Frye, Frye, Daubert) Daubert
Arizona	(1993) 1996, 1997	(Frye) Frye, Frye
California	(1998) 1999, 2003	(Frye) Frye, Frye
Connecticut	(1992, 1994) 1994, 2001	(Frye, Frye) Frye, Frye
Delaware	(1993) 1998	(DRE 702) DRE 702
Idaho	(1990) 1995	(IRE 702) IRE 702
Massachusetts	(1991, 1992, 1993) 1994, 1997, 1997, 1997, 2001	(F, F, F) D, D, D, D
Minnesota	(1989, 1991) 1992, 1993, 1994; PCR: (2002) 2003	(F, F) F, F, F (F) F, F
Mississippi	(1992) 1992, 1996, 1998, 1999, 1999, 2000	(Frye) Frye, Frye, Frye, Frye, Frye, Frye
Nebraska	(1992, 1994) 1997, 1998	(Frye, Frye) Frye, Frye
New Hampshire	(1992) 1994, 2003	(Frye) Daubert, Daubert
Pennsylvania	(1994) 1998	(Frye) Frye
Washington	(1993) 1994, 1995, 1996, 1996, 1997, 2001	(Frye) Frye, Frye, Frye, Frye, Frye, Frye

experienced a change in outcome when they changed from the *Frye* standard to the *Daubert* standard.

This analysis of dynamic jurisdictions suggests four important findings. First, eighteen of the decisions where *Frye* was used to reject DNA evidence occurred in jurisdictions later upholding the science under *Frye* or *Daubert*. This suggests that the legal standard plays a significant role in the development of a scientific jurisprudence, and that legal thresholds significantly affect gatekeeping outcomes. Second, these results demonstrate that a *Frye* analysis certainly makes rejection more likely, at least in initial decisions. Third, within the same jurisdiction, adopting a different legal standard (such as *Daubert*) is correlated with a change in judicial outcomes. Fourth, *Frye* is clearly more likely to result in a rejection than *Daubert* or FRE 702.

Conclusion

These findings imply that the legal model, with its emphasis on legal criteria, is partially correct. The size of the threshold implied by the legal standard does affect the ability of the science to gain clearance or be accepted. In this way, it can be concluded that legal criteria can and do influence judicial outcomes. In terms of political implications, these results also confirm that the standards perform as expected. The *Frye* standard appears to do its job. As a conservative standard, its focus on general acceptance does encourage incremental acceptance of valid and reliable novel scientific techniques. The *Daubert* standard is more liberal, purposely allowing for multiple paths to admissibility. And, indeed, we do see more acceptances with *Daubert*. These findings support the notion that judicial thresholds can and do work to sort science according to the strictness desired by the jurisdiction. These findings also partially explain state-to-state variance in gatekeeping decisions as the result of variance in legal thresholds.

Explaining DNA Gatekeeping:
Democrats, Republicans, and the Northeast

A second clue to judicial behavior suggests that gatekeepers are operating according to their own attitudes toward the science and the politics its represents. The simple expectation is that certain gatekeepers will behave differently because of their political preferences. If the some gatekeepers prefer a science and/or support its advocates, they will be more likely to rule in favor of admissibility. This clue suggests that if preferences vary from jurisdiction to jurisdiction, admissibility might vary as well. In chapter 2 this was conceptualized as an "attitudinal support." Perhaps the most intuitive example is the example of judicial attitudes familiar to most trial lawyers. Trial lawyers know that some judges are just more lenient (or more severe) toward certain litigants. This idea of a friendly (or hostile) court could easily carry over to scientific admissibility, especially when admissibility is coupled with either the defendant or the prosecution advocating a position. Some courts are likely more "prosecutor friendly" or more "defendant friendly." This would certainly affect the admissibility of scientific evidence. In the case of DNA, courts or judges predisposed to favor the prosecution would likely also favor admitting DNA evidence.

There is also a more complex conceptualization of attitudinal support. The complex picture supposes that yes, gatekeepers may have their own attitudes, but they are operating in a political context, such as a particular region of the country, which may constrain those attitudes. For this reason, we cannot be too hasty to assume a direct relationship between judicial preference and gatekeeping outcome. If attitudinal correlations exist, however, someone's (gatekeeper or community) preferences might be at work.

The first cut of the analysis examined gatekeeping decisions for Democratic and Republican state supreme court judges. To test this hypothesis, it was necessary to obtain partisan information on the courts and judges. This was accomplished through a two-step process. First, the names of the judges serving on the cases were obtained. For most courts, these names were found on the opinions. For a few courts, it was necessary to extrapolate unsigned names (estimated judges) from the *BNA* judicial directory or the *Judicial Yellowbook*. Second, most judges were labeled with a partisan label (Democratic or Republican). These labels were obtained either by a direct label (in the *Judicial Yellowbook*) or an indirect label extrapolated from the party of an appointing governor. Third, partisan judges were matched with their "vote" or position (admit or reject) in the DNA case. In terms of judicial votes, 346 were cast by Democrats and 217 by Republicans. Judges whose partisanship was indeterminable (coded as "not available") cast 260 votes.

There are three ways to triangulate the partisanship data, and all three methods disclose a moderate partisan pattern in scientific gatekeeping decisions. The first way to look at the data is to examine court-level information to see if there is a pattern at the jurisdiction level. Most jurisdictions containing "yes" and "no" decisions had constant partisan compositions (Florida, Idaho, Louisiana, Maryland, Massachusetts, Minnesota, New Hampshire, Pennsylvania, Vermont, and Washington). This meant that the ratio of Democrats to Republicans (and N/As) was constant across all decisions in that jurisdiction, seemingly bearing no correlation to the outcome.

One jurisdiction, however, demonstrates an interesting partisan pattern. Alabama had four gatekeeping decisions. In three of these (1991, 1996, and 1998), the Alabama Supreme Court was dominated by Democrats (4 to 1, 5 to 1, and 3 to 3). All three of these decisions ruled DNA evidence inadmissible. In the fourth decision (2002), the Alabama Supreme Court was composed entirely of Republican judges. This decision ruled the DNA evidence admissible. While certainly representing only anecdotal evidence, the relationship between Republicans and support for DNA admissibility is consistent with larger findings.

The second way to triangulate the partisanship data is to look at dissent patterns (see table 3.5) to see if judges of a certain party consistently part company

TABLE 3.5

PARTISANSHIP AND DNA ADMISSIBILITY

Dissenting Opinions	Democrats	Republicans	N/A
We should admit	4	1	9
We should not admit	3	0	6

Judicial Votes	Admissible	Not Admissible
Democrats	77% (265)	23% (81)
Republicans	90% (195)	10% (22)

Chi-square 15.716 (p = .000)
Likelihood Ratio 16.813 (p = .000)
Phi −.167 (p = .000)
One-way ANOVA F 16.109 (p = .000)

with their colleagues when it comes to gatekeeping decisions. An analysis of dissenting opinions revealed twenty-three occasions where judges disagreed with their colleagues about the admissibility of DNA evidence for scientific reasons. Nine dissents wrote against DNA, while fourteen wrote in favor of DNA. There were no "within-court" partisan patterns where a lone Democrat (or a minority of Democrats) would dissent from a court of Republicans or vice versa. It is interesting, however, that no identifiable Republicans dissented against accepting DNA. One Republican dissented in favor of DNA. This means that the dissents rejecting DNA were written exclusively by Democrats and unidentified judges (labeled N/A). Of the seven dissents written by identifiable Democrats, 57 percent were written to protest the acceptance of DNA. According to this data, when Democrats dissent they are voting against DNA the majority of the time. This means the direction of correlation is negative for Democrats.

The third way to triangulate partisanship data is to look at all voting decisions to see which party is more likely to reject or accept DNA evidence. This data compiled all judicial votes in all decisions. A moderate partisan pattern does emerge from a comparison of the votes of these groups of judges (see table 3.5). While 77 percent of Democratic votes supported DNA, 90 percent of Republican votes supported DNA. This means that 23 percent of the Democratic votes rejected DNA while only 10 percent of Republican votes rejected DNA. These findings support a moderate finding of partisanship correlation, where the relationship is positive (supportive) for Republican judges and negative (less supportive) for Democrat judges.

While these partisanship findings are certainly modest, all three examinations of the data yielded similar patterns of correlation. It would appear that Democrats are more reluctant than Republicans to admit DNA evidence. This correlation provides a foothold for an attitudinal model of judicial decision making. Furthermore, if the relationship were proven causal, it seems to provide at least a little information on the type of attitudes informing judicial gatekeeping decisions. Rather than a dimension about new or novel scientific techniques (where one would expect conservatives to exhibit reluctance), the dimension at work appears to be a law-and-order dimension (where one would expect liberals to exhibit reluctance in order to protect the accused). This attitude is fairly reasonable given the fact that defendants overwhelmingly challenge DNA evidence. Liberals are expected to protect the rights of defendants, and Democrats often take a more liberal approach to law and order in modern American politics. Conversely, prosecutors overwhelmingly press for admittance of DNA evidence. Republicans tend to take a conservative approach to law enforcement, and this makes them more likely to support DNA acceptance from an ideological viewpoint. Thus, part of the difference in state supreme court gatekeeping decisions is likely due to ideological differences and variation in the political attitudes of the judges themselves.

Region

Variance in judicial outcomes might also be correlated with regional differences reflected in state supreme courts. The test of attitudinal relationships to judicial outcomes is not only limited to partisan data comparisons. There may also be a regional pattern in gatekeeping decisions due to geographic variation in political preferences. For instance, in U.S. history, the South and the North have traditionally had divergent judicial policy on a host of criminal justice issues. Likewise the West and the Midwest have held different political agendas. It is difficult to hypothesize ahead of time about the way different regions will approach gatekeeping policy. The regions correspond with known differences in political culture and other attitudinal measures, however, which may mean that courts in certain parts of the country consistently approach DNA evidence differently from courts in other regions.

State supreme court decisions were aggregated by region to compare DNA gatekeeping outcomes. Table 3.6 presents a regional breakdown of judicial outcomes. The courts in the Northeast were more than twice as likely (37 percent) to reject DNA evidence as the courts in the Midwest (11 percent), South (13 percent), and West (15 percent), and region is moderately symmetric with judicial outcome in a statistical test. The data were pooled with an indicator for the Northeast, and the results were statistically significant. This means the Northeast significantly departed from other jurisdictions in its likelihood to reject DNA evidence. Thus, the data can be said to support a negative relationship between the Northeast and judicial outcomes admitting DNA. This relationship is tested in multivariate analysis later in the chapter.

A closer look at the DNA decisions in the Northeast provides further insight into this anomaly. Those rejecting decisions were Connecticut (1992, 1994, 1994),

TABLE 3.6

REGION AND DNA GATEKEEPING OUTCOMES

Region	Do Not Admit	Admit
Northeast	10 (36%)	18 (63%)
Midwest	5 (11%)	37 (89%)
West	4 (13%)	27 (87%)
South	8 (15%)	44 (85%)

Likelihood Ratio 6.94 (p = .074)
Phi .227 (p = .048)

Delaware (1993), Maryland (1996), Massachusetts (1991, 1992, 1993), New Hampshire (1992), and Vermont (1995). The Northeast, as a region, accounts for over one-third (37 percent) of those decisions rejecting DNA evidence. By comparison, the South accounts for 29 percent of the rejection decisions, the Midwest accounts for 18 percent, and the West accounts for 15 percent. Comparing just the Northeast and the South, the Northeast decided the majority of its DNA jurisprudence (thirteen out of eighteen decisions) before 1995. By contrast, the South decided less than a quarter of its cases (fifteen out of fifty-two decisions) before 1995. As will be come apparent later in this chapter, the timing of the decision is very important because of the publication of the National Research Council reports in 1992 and in 1996. It may be that the Northeast is an anomaly only because the majority of its cases (72 percent) were decided while the DNA controversy was red-hot.

The attitudinal model, as it relates to region, assumes that attitudes toward the admissibility of novel scientific evidence will vary by region. In this instance, the fact that the Northeast and the South dominate the rejection of DNA is somewhat surprising. The Northeast and the South are not conventionally understood as attitudinal bedfellows. If there is an attitudinal relationship, it would be unlikely for it to be the same for Northeastern courts as for Southern courts. They may be using different attitudinal dimensions to reach the same outcome. For instance, if the Northeast is somewhat more liberal, it might be that the rejection of DNA is related to attitudes toward the rights of the accused in a criminal proceeding. Likewise, if the South is somewhat more conservative, it might be that the rejection of DNA is related to attitudes toward new and novel scientific developments.

Though the data do not allow for a precise attitudinal finding with regard to DNA admissibility and region, they do provide some evidence of a regional difference in gatekeeping jurisprudence. This difference may indicate attitudinal differences or simply timing differences, but it is a fairly blunt measure for a nuanced investigation. Perhaps regional differences in later chapters will illuminate this pattern for gatekeeping variation more generally.

Conclusion

What can we say about the way Democrats and Republicans approach DNA gatekeeping decisions? Well, somewhat cautiously, we can say that Republicans appear more likely to support admissibility than Democrats. Because DNA admissibility is advocated by law enforcement, and because Republicans traditionally support law enforcement, this pattern is not too surprising. As for regional variation, it is interesting to see the Northeast emerge as the region most likely to reject DNA evidence. Little can be concluded from this anomaly with regard to attitudes, however, because timing may be equally important. As a clue to judicial behavior, the role of attitudinal support is still indeterminate in the gatekeeping context; it is clear, however, that attitudinal variance is likely part of the story of state variance.

A Role for Politics

Judicial outcomes for scientific evidence in general, and DNA evidence particularly, are also partly correlated with the political positions of the actors and interests involved. The institutional and organizational advantages can provide political leverage in individual cases. Several political and organizational variables were examined to look at the role of political leverage in state supreme court gatekeeping outcomes. The organization of policy advocates (law enforcement) and the presence of third-party validation (reports by the National Research Council in 1992 and 1996) appeared to have the largest effect on the admissibility of DNA in state supreme court decisions. The following discussion explains the relationship between law enforcement advantages, third-party lobbying, and variation in gatekeeping decisions.

Explaining DNA Gatekeeping: Prosecutorial Advantages

The story of DNA admissibility is partly explained by the institutional and organizational support discussed in chapter 2. DNA evidence comes to individual courts with a certain amount of institutional and organizational support. In every case in this data set, the prosecution supported DNA acceptance and the defense argued for DNA rejection. Hence, the subtitle of this chapter, "Law Enforcement in the Laboratory," is very accurate: DNA was a tool used exclusively for the prosecution against the defendant in every DNA gatekeeping case. And, as the data below demonstrate, on every meaningful measure of institutional and organizational support, the prosecution had the advantage: more success on appeal, more legal expertise (attorneys), more expert witnesses, and more industry support in the form of amici briefs from government and professional organizations. The data also support the conclusion that the magnitude of institutional and organizational support is correlated with gatekeeping outcomes. Furthermore, defense successes are correlated with those instances where the defense had more attorneys or experts in the case. Defense successes are also tied to the institutional and organizational support provided by the 1992 National Research Council report on DNA, as will be discussed in the next section.

Appellant Status

Appellants are assumed to have an advantage in legal proceedings. The party initiating a legal action should have a reasonable expectation of winning. Appeals in criminal law, however, are more routine, especially in capital cases. The story of DNA evidence in state supreme courts tells much about the role of criminal appeals in scientific validation. First, appeals were almost exclusively made by criminal defendants. Of the 153 cases in state supreme courts, only seven were brought by the state (Colorado 2001; Florida 1995; Kentucky 1997; Massachusetts 1992; 1997, Minnesota 1994, 2003). In these cases, the state was appealing an intermediate court's ruling in favor of the defendant. Furthermore, the state only lost two cases. Defendants initiated 95 percent (146) of the cases, and 83 percent of these challenges to DNA were rejected. Overall, the state had a much higher success rate when they were the appellants, winning 71 percent of the cases they appealed. By contrast, defendants only won 17 percent of the cases they appealed.

The 146 cases brought by defendants each represents an instance where the trial court admitted DNA evidence. As petitioner, each appeal is an instance where the defendant argued against the admissibility of the evidence. The routine appeals of criminal convictions allow for DNA evidence to be one issue among many in any given appeal. Something can be learned from this, however, about the nature of judicial acceptance of science. The first is that science will be routinely challenged when it is used in criminal prosecution. This means that gatekeepers will continually be deciding matters of scientific gatekeeping—it will not be a rare item for a court to consider. The second is that, at least with DNA evidence, trial courts overwhelmingly admitted DNA evidence. This suggests that those in support of DNA admissibility were doing a good job convincing judges of its validity and reliability (or vice versa, that those arguing against admissibility had a difficult time making their case). Higher courts will only overturn lower court admissibility rulings on the strictest of terms. Thus, we would expect high court rejections of DNA evidence to be rare.

Counsel Expertise

One source of political power and cognitive persuasion has to do with the skill of the attorneys in the case. Assuming more attorneys represent greater legal expertise on a case, the first cut of the data examined the number of attorneys involved in each case. The state (prosecution) averaged 2.71 attorneys per case. Compare this to the average 1.67 defense attorneys on any given case. This number of attorneys may explain some of the success of DNA in the courts. Overall, the state (prosecution) won 83 percent of DNA admissibility cases. For both the state and the defense, the number of attorneys increased the number of successful cases.

The contrast is especially stark for defense attorneys, however, who saw huge increases in success percentages as the size of the legal staff working on the case increased. For the state, cases with five or more named attorneys enjoyed a 100 percent success rate (while lesser numbers hovered at the 75 to 85 percent success rate). For the defense, cases with one, two, or three named attorneys had a 14 to 16 percent success rate but increased to 50 percent for those cases with four attorneys and 100 percent for the two cases with five defense attorneys. The relationship between the number of defense attorneys and the judicial outcome was modest and negative (phi = −.294, p = .012). This implies that the more attorneys on the defense team, the more likely the court was to reject DNA evidence. This also implies that a portion of state variance may be due to the ability of the defense to mount a successful legal team. It may be that the difference between state policies is partly a factor of the caliber of the defense teams.

This finding supports the hypothesis about organization and political power: the more legal help available to fight the acceptance of DNA, the more likely the court will reject DNA evidence. It is also noteworthy that the increase of legal personnel did not significantly alter the prospects of the prosecution. The number of state's attorneys was not significantly related to judicial outcomes. It may be that DNA rejections required greater legal skill than DNA acceptance, especially when appeals are routine. This implies that all cases were not equal—and thus the outcomes were not equal, even while the science was the same.

The data also presented an opportunity to contrast the success of professional and nonprofessional defense counsel. One might have expected public, paid, professional defense counsel (public defenders, appellate defense counsels, etc.) to have more expertise in these cases. The data, however, demonstrate that private counsel (either retained or appointed) enjoyed a 20 percent success rate; compare this to 14 percent success among public defenders. This coincides with the descriptive finding that private defense attorneys win more cases than public defenders. This may be a function of workload and resources. Legal scholars have long recognized the advantage of private defense counsel. In the scientific context, one might feel that public defenders would have more expertise battling routine DNA admittance. The data, however, support the advantage of resources and skill over pure experience.

Expert Testimony

Another source of political power and cognitive persuasion has to do with the quality of expert testimony. Supporting experts should add to the success of judicial outcomes because they assist the court in making its policy decision. Assuming that more experts represent more persuasive power, the first cut of the data examined the number of experts presented by the defense and the prosecution. The prosecution averaged 1.05 experts per case. The defense averaged .29 experts per case.

The second cut of the analysis examined the type of expert employed by each side (see table 3.7). Experts for DNA decisions generally fell into three

TABLE 3.7

TYPES OF EXPERTS IN DNA CASES

Type of Expert	Defense		Prosecution	
	TOTAL CASES	TOTAL PEOPLE	TOTAL CASES	TOTAL PEOPLE
Lab	2	2 (5%)	72	86 (54%)
Law enforcement	1	1 (2%)	27	29 (18%)
Academic	25	41 (93%)	28	45 (28%)
Total		44		160

TABLE 3.8

ROLE OF DEFENSE EXPERTS IN DNA CASES

	Cases	Admitted	Rejected
Cases with zero defense experts	126	107 (84%)	19 (16%)
Cases with one or more defense experts	27	19 (70%)	8 (30%)

categories: lab personnel, law officers, and academics. As a proximate cue for a human decision maker, it is not clear if "insiders" (lab experts and law enforcement experts) would carry more weight than "outsiders" (academics removed from the legal system). Insiders may do a better job of addressing the specific legal needs due to their experience with criminal justice jurisprudence. Outsiders, however, may be viewed as more trustworthy as a result of their apparent neutrality. What is clear is that defendants had a hard time finding or employing "insider" experts. In all, 93 percent of defense experts were academics. Compare this to the 28 percent of state experts who were academics. The state clearly favored using lab experts, as 54 percent of expert testimony originated from the lab.

A third type of analysis looked at the role of law enforcement experts in these cases. One interesting finding is that 31 percent (49 out of 153) of the cases involved labs operated by law enforcement agencies. These labs supplied 56 of the 85 lab experts used by the state. When these experts were combined with other law enforcement (non-lab) personnel who provided expert testimony, we see that 53 percent of state experts were employed by law enforcement agencies.

A fourth look at the analysis examined the role of defense experts in judicial outcomes (see table 3.8). In twenty-seven cases, the defense used expert testimony. (While more experts may have been at the trial, this measure captures only those experts mentioned by name or occupation in the high court opinion.)

(Information about trial-level experts not mentioned in these opinions are impossible to obtain.) Comparing these cases to the 127 cases where no defense experts are mentioned supplies some support for the added value of expert testimony. The defense experienced 30 percent success with an expert, as opposed to 16 percent success without an expert. Thus, the variation in state supreme court gatekeeping policy is partially the result of differential use of expert testimony. Those courts hearing from a host of defense experts would certainly be more likely to give more consideration to DNA rejection.

Overall, this analysis of expert testimony demonstrates a political and organizational advantage for the supporters of DNA evidence. This political leverage can certainly persuade (or dissuade) courts of the reliability and validity of DNA analysis. State prosecutors were able to secure more experts (including "insider" experts) to testify on behalf of DNA. Defendants found it difficult to make use of expert testimony. This could be the result of poor attorney training, fewer resources for the retention of experts, or the availability of experts to criticize DNA analysis. Whatever the cause, expert testimony provides significant political leverage in individual gatekeeping decisions.

Lobbying the Gatekeepers

Political forces outside the immediate parties to a case may also significantly affect state supreme court gatekeeping policy. Two sources of organizational political leverage particularly important for DNA gatekeeping decisions are policy advocates and third-party reports. Policy advocates are interested third parties who lobby a particular supreme court on a particular case. Third-party reports are more generally available (or unavailable) in the policy environment and may be used by attorneys to make their case to a court. The variation in state supreme court DNA decisions is at least partially explained by organizational political leverage operating on certain decisions. The following analysis examines the role of amici briefs and the National Research Council in DNA gatekeeping decisions.

Policy Advocates

Policy advocates may formally lobby courts by submitting amici curiae, or "friend of the court," briefs to provide additional information or arguments to courts considering cases of interest to them. A surprising number of decisions recorded amicus briefs for DNA cases in state supreme courts (see table 3.9). Nine decisions recorded a total of twenty amici briefs. While too small to draw statistical conclusions about the importance of amici in scientific gatekeeping decisions, this sample is still informative. First, there appear to be three types of policy advocates in these decisions. One type of policy advocate is internal to the legal system: public defenders and public prosecutors. The fact that four briefs represented public defenders and nine briefs represented district attorneys, attorney generals, or city

TABLE 3.9

AMICUS BRIEFS IN DNA CASES

Decision	(Admit?)	Amici for Defendant	Amici for State
California 1998*	(No)	-Two private attorneys	-District Attorney
California 1999	(Yes)	-Public Defender's Assoc.	
Colorado 2001	(Yes)		-Attorney General
Minnesota 1994	(Yes)		-Attorney General -Institute of Human Genetics and Mangen Research Assoc. Inc. -Minnesota City Attorney's Assoc.
Missouri 1996	(Yes)	-National Legal Aid and Defender's Association -Two professors of law	
New York 1994*	(Yes)	-Criminal Law Clinic	-District Attorneys Assoc. of NY -Atty. Gen. for Div. of Crim. Justice
Ohio 1992*	(Yes)		-Attorney General
Washington 1993*	(No)	-Whatcom Co. Public Def.	-Attorney General -Cellmark Diagnostics
Washington 1995	(Yes)	-Am. Civil Liberties Union -Seattle/King County Public Defender's Association	-Attorney General -Nat. Assoc. of Parents of Murdered Children

*First high court DNA case for the jurisdiction

attorneys is evidence of strong policy recommendations by legal personnel. Briefs representing law enforcement and criminal defense provide a picture of the values at stake in the use of this particular science. A second type of policy advocate is the industry supported by the science. Two briefs were filed by laboratories that specialize in DNA analysis: Cellmark Diagnostics Inc. and the Institute of Human

Genetics. As expected, industry is supportive of the use of DNA in criminal prosecution. A third type of policy advocate is external to the legal system: public-interest groups. The American Civil Liberties Union and the National Association for Parents of Murdered Children both filed amicus briefs. These groups represent the conflicting values in the larger political context, namely, the desire to protect the rights of the accused and the desire to bring the guilty to justice. The presence of two other amicus briefs by unassociated attorneys and law professors (both in support of the defendant) may or may not mean that there is a fourth type of policy advocate (legal experts). Because the nature of the briefs cannot be analyzed, little commentary can be made on this group. It is simply noted that other authors of briefs existed.

Does policy advocacy matter? There is little to be drawn from these modest findings. There is no clear pattern relating the presence of amici to judicial outcomes except to say that eight of the ten decisions ruled in favor of the state, and two of those decisions did not have amici supporting the state. In a broader policy context, it is clear these provide cues to judges about the values and resources at stake in scientific gatekeeping decisions. Legal personnel, industry, and public-interest groups, as well as legal experts, clearly care about judicial gatekeeping outcomes. It is not surprising to see that the majority of the briefs represent advocates who have the most at stake in gatekeeping decisions: the attorneys who represent the accused and the attorneys who are charged with prosecution. As for explaining the overall patterns of acceptance of DNA evidence, there is no clear conclusion about the role of policy advocates. The only finding is that groups with political power represented both acceptance and rejection. The other finding is that the majority of amici cases occurred prior to the 1996 National Research Council report. As will be discussed below, this report reduced significantly the DNA controversy.

DNA Gatekeeping and the National Research Council

Another source of institutional and organizational support is the availability of third-party reports on the acceptability of the evidence for legal purposes. Two key reports were published by the National Research Council (NRC) on the reliability and validity of DNA fingerprinting. These reports are an important part of the DNA gatekeeping story. A report with "legitimacy" certainly aids the gatekeepers and advocates in their task. A third-party report also takes political organization to commission and fund such reports. Government and law enforcement officials, especially the FBI, pushed for NRC studies to offer "impartial" and expert analysis of DNA fingerprinting for criminal prosecution.

The first NRC report (hereinafter NRC 1992 report) criticized only the statistical part of DNA analysis, assuming proper lab protocols were maintained. The NRC 1992 report undertook an extensive examination of DNA testing and legal issues. The report emphasized that the use of DNA fingerprinting for criminal prosecution required reliable methods for all steps in DNA analysis, including the final step of generating reliable statistics. With regard to population frequency statistics, the research council concluded that the use of theoretical models, rather than actual counts in a real population, made the statistics highly unreliable. In particular, the assumptions of the product rule were not sufficiently tested in the real human population to assume that a match at loci one was unrelated to a match at loci two (and so on). The worry is that a DNA match between unrelated persons (for instance an innocent defendant and a perpetrator) might be more likely than the statistics assumed—thus resulting in the miscalculation of the likelihood of the crime-scene DNA matching the defendant. In lay terms, there was also a worry that certain populations, such as Native Americans, might share several DNA markers, and thus the arrest of any local Native American would result in a "match" possibly convicting the wrong person. To combat this error, the research council recommended the use of a conservative estimation statistic (the ceiling principle). They also explicitly declared the use of any other estimation statistic, especially the product rule, unreliable and invalid.

The NRC 1992 report had the effect of giving defense attorneys a powerful tool appealing DNA trial evidence. Defense attorneys could argue the prosecution wrongly used the product rule when giving frequency statistics to the jury. Courts could also reach the conclusion that there was no "generally accepted" way to calculate frequency statistics—leading to a rejection of DNA evidence altogether. This was reasonable given the research council's insistence that all steps in the DNA typing process must be proven reliable and valid for DNA to be acceptable scientific evidence in a criminal trial.

The second NRC report (NRC 1996 report) amended the statistical concerns and effectively closed the DNA analysis debate among the expert community. In the NRC 1996 report, the research council revisited the question of population statistics. In four years an impressive amount of research in population genetics and research, as well as DNA typing more generally, had been completed. The research council examined the latest state-of-the-art measurements and concluded that "much has been learned since the last report" (1). The bottom line in the NRC 1996 report was an emphatic endorsement of DNA typing techniques and population frequency testimony: "The technology for DNA profiling and the methods for estimating frequencies and related statistics have progressed to the point where the reliability and validity of properly collected and analyzed DNA data should not be in doubt" (1). While the 1992 report had the effect of aiding both the supporters and nonsupporters, the 1996 report essentially gave a green light to those in favor of DNA admissibility.

These reports conceivably supplied judges with additional judgment material, and it is reasonable to suspect they have some bearing on judicial outcomes with regard to DNA. When it comes to sorting the role of the NRC reports in state supreme court variation, there are two ways to approach the analysis. One approach is to focus on cases occurring after the reports were issued and compare outcomes for those decisions citing the NRC reports with those who did not. A second approach is to assume that the reports were part of the policy context after their publication and simply compare cases occurring before and after publication.

The first cut is to compare cases discussing NRC reports to those that did not (see table 3.10). For citations of the NRC 1992 report, we must look only at state supreme court decisions occurring after publication. In DNA cases occurring from 1992 to 2003, 34 percent (46 cases) mentioned the 1992 NRC report. Of these cases, 89 percent (41 cases) followed NRC recommendations. Indeed, the citation of the 1992 NRC report is moderately related to a negative judicial outcome for those cases occurring after 1992.

Turning our attention to the 1996 NRC report, a relationship between the report and judicial consideration also emerges (see also table 3.10). For cases from 1996 to 2003, 25 percent (20 cases) mentioned the 1996 NRC report. Of these cases, 100 percent (all 20) followed NRC recommendations. By comparing cases occurring before (24 percent rejected) and after 1996 (8 percent rejected), one can see that there is a weak relationship between the date and the outcome. By comparing cases after 1996 that cited (0 percent rejected) and did not cite (13 percent rejected), one can see that there is a weak relationship but it is not enough to reject the null hypothesis due to the small number of cases for comparison.

Because the report is present in the decision universe, it also makes sense to step back from citation and simply compare those cases decided before publication with those cases decided after publication. Here, for cases occurring from 1989 to 1996, no significant relationship is found between the decision date and

TABLE 3.10

THIRD-PARTY REPORTS AND DNA ADMISSIBILITY

1992–2003*	Cases	Admitted	Rejected
NRC 1992 cited	46 cases	33	13
NRC 1992 not cited	75 cases	69	6

Phi = −.27 (p = .003)
*National Research Council issued a report on DNA evidence in 1992

1996–2003*	Cases	Admitted	Rejected
NRC 1996 Report cited	20 cases	20	0
NRC 1996 Report not cited	60 cases	52	8

Phi = −.196 (p = .015)
*The National Research Council issued an updated report on DNA evidence in 1996.

the judicial outcome. In all, 25 percent of cases occurring before 1992 rejected DNA evidence and 23 percent of cases after 1992 rejected DNA evidence. Even for the entire span of years (1989–2003), the occurrence of the case before or after the publication of the 1992 NRC report is not significant. This is perhaps due to the inconclusive nature of the NRC 1992 report. It did not effectively resolve an issue. In addition, several states had already admitted DNA evidence and did not meaningfully revisit their decision.

Because citation of the NRC 1992 report was related to judicial outcomes, further analysis created a three-part comparison: those cases occurring before 1992, those cases occurring after 1992 that cited the NRC report, and those cases occurring after 1992 that did not cite the NRC report. Of those occurring before 1992, 25 percent rejected DNA evidence. After 1992, rejection rates differed appreciably between those cases citing (26 percent rejected) and not citing (9 percent rejected) the 1992 NRC report. Overall, these three groups of cases have a moderate relationship with judicial outcomes (phi = .220, p = .025). Taken together, this analysis of the 1992 NRC report and judicial outcomes suggests that the NRC report was an important part of the judicial consideration of DNA evidence for some jurisdictions. It also signifies that many of the decisions to reject DNA evidence could be due to the fact that the citation of the NRC report had a negative effect on judicial admissibility. Likely, this was due to those cases in the analysis where the court used the 1992 NRC's skepticism of the statistical methods used in match testimony.

It is somewhat helpful to look more generally at "rejection" patterns after 1996, as that is when the second report effectively silenced most debate. As mentioned earlier, state court rejections noticeably taper off after 1996 (revisiting table 3.1). As a matter of fact, 70 percent (19 cases) of rejections occurred before 1996. Table 3.11 compares acceptance patterns before and after 1996. According to these cases, 63 percent of favorable rulings occurred after 1995. Even more interesting, thirteen states waited until 1996 or later to rule on the admissibility of DNA. Only two of these states (California and Louisiana) ruled against DNA, and both of them later accepted DNA.

TABLE 3.11

DNA COURT ACTIVITY BEFORE AND AFTER 1996

	Admit DNA	Reject	First DNA Decision
1989–1995	46 cases (37%)	19 cases	35 states
1996–2003	80 cases (63%)	8 cases	13 states

It is also helpful to look at the time period when dynamic jurisdictions (refer back to table 3.2) shifted in their DNA jurisprudence. Most jurisdictions show a shift corresponding with either 1992 (Idaho, Minnesota, New Hampshire) or 1996 (Arizona, Delaware, Nebraska, Pennsylvania). These shifts may or may not be evidence of the important role of third-party validation in judicial outcomes. There is a pattern of acceptance falling within the time periods associated with the NRC reports, however, and these reports likely played a role in DNA gatekeeping jurisprudence.

Multivariate Analysis of DNA Gatekeeping Decisions

By placing likely variables together in a multivariate analysis it is possible to test for independent and significant effects. The most informative model (see table 3.12) included three of the most important variables: the legal standard, whether the case was in the Northeast, and whether the case was before or after 1996. The dependent variable in this analysis is the judicial outcome for each case, coded as a binary dummy variable: zero for "not admitted" and one for "admitted." Data collection coded on thirty-eight separate variables. Of these variables, several reflect theoretical concerns: legal model (legal standard), attitudinal model

TABLE 3.12
LOGIT ANALYSIS OF JUDICIAL OUTCOMES FOR DNA EVIDENCE

	B	S.E.	Wald	df	Sig.	Exp(B)
Legal standard	.689	.338	4.147	1	.042	1.991
Region is not NE	1.257	.501	6.298	1	.012	3.517
Before 1996 NRC	−1.119	.550	4.142	1	.042	.327
Constant	.349	.767	.207	1	.649	1.418

Chi-square (significance) 18.947 (.000)
−2 Log Likelihood 123.648
Nagelkerke R-Square .192
$N = 153$

Prediction Table		Predicted		Percentage Correct
		ADMIT	DO NOT ADMIT	
Observed	ADMITTED	121	5	96.0
	NOT ADMITTED	19	8	29.6
				84.3 overall percent

(region), political/organizational model (defense experts, state experts, NRC 1992, NRC 1996, defense attorneys, public defenders, and state attorneys). Table 3.12 presents the best statistical model. The legal model variable was coded by assigning a number to each standard: *Frye* general acceptance (1), *Daubert* (2), FRE 702 (3), and "other" (4). The region variable was measured categorically using the classifications used by the U.S. Census and collapsed into a simple Northeast indicator. Cases were coded for whether they occurred in the Northeast. A third measure coded the presence or absence of citation of NRC reports for 1992 or 1996. This variable was measured several ways as described above in reference to third-party reports. Because the bivariate analysis revealed a relationship for the 1992 NRC report and the 1996 NRC report, several models were developed to test each of these third-party validations separately.

Overall, the chi-square of 18.947 (.000) allows us to reject the null hypothesis (that all the coefficients in the model are zero). Here, the model performs better than the baseline, by correctly classifying 84.5 percent of the cases. Even with such a small percentage of "rejections" this model is able to find contributions from three variables: the legal standard, region, and the cases relationship to the 1996 NRC report (coded as before or after 1996). According to the estimated R-square, this model accounts for 19 percent of the variance.

Multivariate Discussion

In terms of specific variables, this model supports several hypotheses. First, the legal standard is significantly related to the outcome, even with the other variables held constant. As expected from the bivariate analysis of *Frye*, *Daubert*, and FRE 702, legal standard has an independent effect even when region and the other variables are held constant. The model contains a positive coefficient for the legal standard variable. The fact that the coefficient is positive indicates that judicial acceptance increases as the standard's liberality increases.

Second, the region of the case has a notable significant effect as well. Whether the case originates in the Northeast is related to the judicial outcome, even when the legal standard and other variables are held constant.

Third, the presence of third-party reports had mixed results. It is not surprising to find less statistical support for the 1992 report than the 1996 report. The 1992 NRC report was supportive of much of the use of DNA evidence, yet it questioned the use of certain probability statistics for DNA analysis. While some cases used the report to bolster a finding of admissibility, several cases rejected DNA evidence because of statistical questions. After the 1992 report was available, thirteen of the forty-six cases (28 percent) citing the NRC rejected DNA evidence. Compare this to eight of the seventy-five cases (8 percent) where the NRC report was not cited and the DNA evidence was rejected. Clearly, DNA fares worse in cases where the 1992 NRC report was cited. Yet another way to look at the data is to look at the qualitative story. Before 1992, defense attorneys had little means to challenge DNA evidence, especially statistics. The NRC report provided at least one avenue for challenge. Again, a look at table 3.1 demonstrates that fourteen of the twenty-seven successful challenges to DNA evidence took place between 1992 and 1995. Before 1992, few cases were rejecting DNA evidence. By contrast, this model indicates that the location of the case before or after 1996 was significantly related to the judicial outcome and in the expected direction. Cases before 1996 were negatively related to admissibility, even when region and legal standard were held constant.

Clues to Gatekeeping Behavior

This quantitative analysis has uncovered several correlates of judicial gatekeeping decisions. The theoretical foundation of this study was built on the concept of judges as human decision makers who frame and evaluate information. This cognitive process was subject to three levels of influence. One level is the role of legal standards in framing the analysis for the decision maker. Another level is the role of the decision maker's own attitudes in framing the analysis. A third level is the role of external factors on the framing of the analysis.

The findings in this study found a significant role for legal standards in the explanation of state supreme court variance in DNA gatekeeping policy. Indeed legal standards were significantly related to case outcome, both in correlation and in higher-level regression analysis. The role of standards in framing the question for courts clearly has the expected result of altering judicial outcomes. The more liberal the standard, the more likely the evidence was to be accepted.

As for attitudes, partisanship was moderately correlated with judicial behavior, possibly providing an additional explanation for state variance. The findings suggest a role for "law-and-order" attitudes in judicial gatekeeping decisions. Slightly partisan behavior was observed in gatekeeping decisions, with Democrats more likely than Republicans to reject DNA evidence. It was also found that courts in the Northeast and in the South were more likely to reject DNA evidence than courts in other regions of the country. This too may indicate an attitudinal relationship to judicial outcomes, for this relationship held even when third-party validation and legal standard were held constant. This implies that states or courts will vary in their DNA gatekeeping policy as a result of their political attitudes.

In terms of political positioning, the evidence also recognizes the role of institutional and organizational support in DNA jurisprudence and state variance. Third-party reports, at least in timing and citation, found some relationship to judicial outcomes. While not significant in multivariate analysis, timing of decisions and reports appeared related. The majority of rejections occurred before 1996. Likewise, citations of the 1992 NRC report tended to decrease judicial acceptance as evidenced by numerical comparison and a qualitative investigation of what courts were saying when they cited the report.

The Political Implications of the Findings

The first political implication has to do with the use of legal standards to determine which science should and should not be part of legal proceedings. The data on DNA evidence support the conclusion that these standards function well. More liberal standards will more quickly accept novel scientific techniques. For those jurisdictions and constituencies who prefer this public policy, they can rest assured that a liberal standard works. Similarly, more conservative standards will slow down the acceptance of novel scientific techniques—or possibly preclude acceptance altogether. For those jurisdictions and constituencies who prefer this more incremental public policy, they are getting the desired outcome. The data on DNA evidence also supports a conclusion that these standards do control judicial outcomes and account for state court policy variance. In this sense, law matters in judicial politics.

The second implication regards the role of political attitudes and partisanship in judicial outcomes. The finding of a moderate partisanship and regional pattern implies that judges may admit or reject science with an eye toward its policy implications. Thus, the political implications of a particular science will affect its movement through the judicial system because the gatekeepers use political attitudes in their decision making.

The third implication regards the role of political and organizational resources in the judicial processing of scientific information. DNA evidence is overwhelmingly used by state law enforcement to prosecute defendants. This means that all the organizational and political power of law enforcement institutions can be brought to bear in the battle for acceptance. Laboratories and expert witnesses, as well as legal personnel, will always provide the state with an advantage in forensic science. Furthermore, law enforcement organizations (such as the FBI) can use their political strength to push for third-party reports. The next chapter will further test this law enforcement hypothesis by presenting a type of scientific evidence sought by both the prosecution and the defense: polygraph evidence.

CHAPTER 4

LIE DETECTION

VICTIM OF LAW AND POLITICS

The judicial debate over lie detectors, or polygraphs, has been very long and involved. The original U.S. Supreme Court decision on the admissibility of scientific evidence, *Frye v. United States*, 293 F. 1013 (1923), was an early polygraph case. In that case the Court announced the "general acceptance" test for scientific evidence. It also held that the polygraph examination had not yet attained "general acceptance in the relevant scientific community." Since that time, 163 state supreme court cases have revisited the scientific reliability and validity of polygraph evidence. Twenty-nine decisions rule polygraph evidence admissible, while the remainder rejected polygraph evidence for criminal trial proceedings. Here we are faced with a mystery similar to that of DNA admissibility: why did some jurisdictions admit while most rejected? Presumably the science of polygraphs was quite similar case to case, yet different states arrived at different conclusions as to the reliability and validity of lie detector evidence in criminal trials.

Polygraph evidence is less straightforward than DNA evidence—which was itself hotly contested. For DNA, the debate was largely centered on the technique and the statistics used to generate match frequencies. With polygraph evidence, the controversy increases for two important reasons. One has to do with the purpose of the science. Detecting untruthful behavior is legally different than identifying material from a crime scene because the jury—not science—is supposed to assess witness veracity. The other has to do with the role of human interpretation in polygraph assessment. Polygraph evidence presents a multidimensional legal question and is subject to various levels of analysis, leading to a multiplicity of court responses.

The Science of Detecting Lies

The science of polygraph evidence involves the triangulation of many measurements of human physiological response. The theory of polygraph evidence is rooted in a belief that psychology affects physiology. In other words, psychological states can be manifested in detectable physical responses. Deception or fear of detection (psychology) is manifested in heart rate, blood pressure, respiration, and electrodermal response (physiology). Where DNA evidence had three types of science (theory, technique, and statistics), polygraph adds special problems with indirect measurement and subjective interpretation.

Polygraph results are generated by a three-step process. First, the polygrapher calibrates the machine during a pretest interview with the respondent. The respondent is monitored with a cardiograph (heart rate), sphymograph (blood pressure), pnuemograph (respiration), and galvanometer (electrodermal response). The pretest interview demonstrates to the respondent that the machine can tell a difference in responses to neutral, control, and relevant questions. Second, the actual test records the respondent's physiological data during the interview. Third, the physiological responses to relevant and neutral questions are compared and scored by the polygrapher. The result of the polygraph is usually described as "indicating deception" or "indicating truthfulness."

Theory

Two assumptions form the foundation for polygraph evidence. The first assumption is that of a relationship between deception and an individual's emotional state. The second assumption is that there is a relationship between the emotional state and certain measurable physiological responses (heart rate, blood pressure, respiration, electrodermal response). Thus, the theory of the polygraph is that measuring physiological responses is in effect the measurement of the individual's emotional state. Furthermore, the theory assumes an emotional state will be different for truthful versus deceptive individuals because deceptive individuals will fear detection. This fear of detection creates an emotional state of anxiety, which can be measured using the physiological data.

The primary challenge to these assumptions arises from the problem of indirect measurement. The physiological measurements are not unique to fear of detection. Several concerns unrelated to deception may create an emotional state of anxiety, thus increasing the physiological response for a reason unrelated to deception. Polygraph practitioners have countered this concern with observation that it is the *changes* in response to different questions that indicates deception. Thus, a person with an anxious emotional state unrelated to deception would exhibit constant physiological readings. In other words, the anxious innocent individual would have a high heart rate for both neutral and relevant questions.

A secondary challenge to this theory is the fact that indirect measurement of a fear of detection is subject to intentional manipulation. A deceptive subject could use countermeasures to foil the measurements. One example is the use of pain to create artificial physiological response to control and neutral questions, thus leading the data to record no increased anxiety for the relevant questions.[1] Polygraph examiners have suggested that countermeasures can be overcome with proper policing of the polygraph exam.

Of course, the problem for courts is to decide when or if this theory has any validity for criminal trial proceedings. In reference to the polygraph, the *Frye* Court stated the problem this way: "Just when a scientific principle or discovery crosses the line between the experimental and demonstrable stages is difficult to define. Somewhere in this twilight zone the evidential force of the principle must be recognized" (*Frye v. United States*, 293 F 1013, 1014 [1923]).

Courts have generally had two responses to polygraph evidence. One response is to accept the "theory" while rejecting the technique: "It is based on scientific theory, but remains an art with an unusual amount of responsibility placed on the examiner" (*People v. Anderson*, 637 P.2d 354, 360 [Colo. 1981]). Another response is to dismiss the theory itself as unsubstantiated: "The physiological and psychological bases for the polygraph examination have not been sufficiently established to assure the validity or reliability of the test" (*State v. Fain* 774 P.2d 252, 257 [Idaho 1989]).

Judges are further confronted with empirical evidence of widespread use of polygraph machines as they wrestle with discerning the general acceptance of the theory. Polygraphs are routine in private and government employment, as well as criminal investigation. This level of utilization is at least enough for most courts to concede general acceptance of the theory: "As a matter of common knowledge, polygraph evidence has developed to the point that it is used in industry's determination of hiring and firing, in law enforcement, by national security agencies such as the Central Intelligence Agency, the Armed Forces Intelligence agencies, and the Federal Bureau of Investigation" (*State v. Harrod*, 26 P.3d 492 [Ariz. 2001]).

Technique

The controversy over the polygraph exam does not arise from the use of technical instruments. With regard to the measurement of physiological responses, "There is no question that a high quality polygraph is capable of accurately measuring the relevant physical characteristics" (U.S. Congress, *Office of Technology Assessment Report on Polygraphs* 1983). Controversy arises in the translation of those measurements into an assessment of deception. This is ultimately the responsibility of the polygraph examiner, and it involves subjective interpretation.

The examiner is active at three crucial steps in a polygraph exam. First, the examiner designs and implements the test. This involves several tasks. The examiner calibrates the machine during the pretest. The pretest interview is designed to convince the subject that the machine can detect deception. This should agitate the guilty subject while calming the innocent subject. This is accomplished using a "stimulation test." Here, the subject chooses a card or number, known only to him or her. The examiner asks for each alternative, instructing the subject to reply, "No," to all alternatives, even the one chosen by the subject. From this test, the physiological data should allow the examiner to deduce the "correct" alternative, thus convincing the subject of the reliability of the exam.

Second, during the actual test, three methods of questioning may be used. The "control" test uses control, neutral, and relevant questions. The "relevant-irrelevant" test uses only relevant and irrelevant questions. The "guilty knowledge" test assumes the subject will have a greater response to the correct alternative on a list of informational items taken from the crime scene. All three of these tests are necessarily designed to be case specific, and this gives the examiner broad latitude in test design.

Third, the examiner scores the exam by comparing physiological data for irrelevant answers to data for relevant answers. Commonly this is done using a +3 to −3 scale with 0 meaning there is little difference. A score of +3 indicates little response to the relevant question. A score of −3 indicates a strong response to the relevant question. These scores are then averaged for a decision of truthful, deceptive, or inconclusive. More recently, computerized scoring programs have been developed.[2] This is an attempt to overcome the subjective nature of examiner scoring. The few courts ruling on computer-scored polygraphs, however, have not been favorable. For example, Kansas State Supreme Court noted expert evidence, which "cautioned that all aspects of the computerized psycho physiological detection of deception are still in the developmental stages, necessitating cautious scrutiny" (*State v. Shivley*, 999 P.2d 952 [Kan. 2000], reviewing commentary by the director of the Department of Defense Polygraph Institute appearing in the *Journal of Forensic Science*, 959).

The Challenge of Lie Detection

Courts overwhelmingly express concern about the primary role of the polygraph examiner in the process. Unlike DNA testing, subjective human impressions are likely to contaminate the pretest calibration, the interview itself, and the scoring. The interpretation of the test is the crucial step in the process, linking the physiological data with a determination of deception or truthfulness. The fact that individual examiners could arrive at a different result given the same subject is cause for alarm and distrust.

Proving That Results Are Valid

The first challenge of lie detection is proving that results are valid. In judicial proceedings, courts like to hear information on the validity of results. With DNA evidence, this was generally referred to as statistical analysis. With regard to polygraph evidence, proving the validity of results is more nuanced than a simple statement of the likelihood of a random match. From a scientific standpoint, the problem of gauging the validity of results is probably the most significant bar to polygraph admissibility. Validity questions are present in two forms.[3]

The first question is how do you know when the test is accurate enough? Accuracy about the ability to detect deception by a guilty person (referred to as sensitivity of the test) is as important as the ability to detect truthfulness by an innocent person (referred to as the specificity of the test). For instance, if those who are in fact guilty are labeled "guilty" 90 percent of the time, is that accurate enough? Conversely, if those who are in fact innocent are labeled "innocent" 80 percent of the time, is that accurate enough? Polygraph examiners have not been able to agree on the degree of sensitivity and specificity required to label an "exam" good.

Polygraph supporters have tried to use both field tests and lab experiments to determine the accuracy of the polygraph technique. In the 1960s a prominent publication touted a 70 percent accuracy rating. In the 1980s a different publication proclaimed polygraph tests were 95 percent accurate (Raskin 1986). Critics cannot agree on which polygraph experiments are valid measures of the actual accuracy of the test. They are particularly skeptical of lab experiments because there is no way to simulate the normal anxiety of the suspect situation or the stakes inherent in a police interview (Giannelli and Imwinkelried 1993).

The second question concerns probability. Given a "deceptive" reading, how likely is it that the person is truly deceptive? The predictive values of polygraph evidence have not been sufficiently established. This lack of probability of accuracy is clearly a concern and a problem difficult to overcome. Due to the difficulty of finding out how many deceptive readings are actually wrong, there are inherent difficulties in obtaining "real world" data that is completely valid.

Special Legal Concerns

A second challenge unique to lie detection has to do with special legal concerns. Courts are quick to find that the unique position of polygraph evidence in the legal system raises the bar for admissibility. Unlike DNA evidence, polygraph tests threaten the legal system in a particular way. Unlike DNA, polygraph evidence invades the fact-finding province of the jury. In our legal system, juries are given the sole task of ascertaining the veracity of witnesses. The polygraph exam threatens to usurp that solemn position by "telling" the jury which witnesses are being deceptive. DNA evidence spoke only to the likelihood of identity. Polygraph evidence speaks to the truthfulness of a stated fact: "We are acutely aware that the polygraph is unlike other scientific evidence, since what it attempts to measure—the truthfulness of the witness or defendant—is so directly related to the essence of the trial process" (*People v. Anderson*, 637 P.2d 354 [Colo. 1981]).

Polygraph evidence also increases the possibility for a "battle of the experts." The tests and the subjective role of the examiner greatly increase the parameters for debate. While DNA is a relatively direct measure of loci matches, the polygraph is an indirect measure of an emotional state, with the added problem of subjective human interpretation. Some commentators refer to the problem of the "friendly polygrapher" as exacerbating this difficulty (Giannelli and Imwinkelried 1993, 230–233). This theory hypothesizes that examinations privately conducted by the defense are unreliable. A defendant may not have a sufficient level of fear of detection to make the test reliable (Orne 1975). For this reason, courts are significantly challenged to decide when, if ever, to accept polygraph testing in criminal proceedings. As the data below indicate, most courts found that these challenges closed the gates on polygraph evidence. Other courts, however, arrived at a very different conclusion regarding the reliability and validity of polygraph evidence.

The Politics of Lie Detection

As a gatekeeping task, the admissibility of polygraph evidence essentially involves a judgment about two ways of knowing: the theory of physiological response indicating psychological reality and the application of polygrapher knowledge and assessment in the testing and scoring of a subject. No court has a problem with the actual physiological measurements. Some courts express concern over the theory. Most courts have a problem with the subjective nature of polygrapher-controlled results.

The DNA analysis in the last chapter focused on the role of three political processes in judicial outcomes: the process creating resources for a legal contest, the process creating the calculus for legal decision making, and the cognitive framing process used by the decision makers themselves. As with DNA, law enforcement and government heavily controls the proliferation of polygraph availability. The polygraphy industry has been driven by the demands of government and law enforcement officials. It has its own journals and professional associations. Polygraph institutes are prevalent, and polygraph training is standard for many law officers.

The breadth of this institutional and organizational support might presuppose impressive acceptance rates when polygraph evidence is presented in courts. Unlike DNA, however, polygraph evidence serves its primary prosecutorial purpose *outside* judicial proceedings. DNA is often central to linking a perpetrator to a crime scene, and DNA admissibility is of primary importance for prosecutorial purposes of obtaining a conviction. In contrast, polygraph admissibility at trial is a secondary concern in prosecution strategy. Polygraphs perform their most important tasks during the investigative phase of the prosecution, where they are used in two important ways. First, they are used to sort suspects and witnesses. Second, they are used to induce confessions. The ability to present polygraph results to a jury is a secondary concern. By contrast, while DNA may be used for investigative purposes, it is often a crucial piece of prosecution evidence in the courtroom.

Prosecutors and law enforcement establishments have not prioritized acceptance of polygraph in the courtroom. Indeed, in most of the cases in this analysis, the defendant is the one supporting the admittance of polygraph evidence. Thus, the political and organizational advantage of law enforcement does not lie in incentives to procure judicial approval. In terms of political leverage, the defendants pushing for acceptance are unorganized and enjoy little political power because of their diffusion. Thus, while law enforcement fought for DNA admissibility, there is a much less political leverage at work for polygraph admissibility.

Defendants also have incentives that keep them from mounting significant assaults to procure judicial approval. Defendants have limited resources with which to seek their acquittal. There are multiple paths to acquittal, of which favorable polygraph evidence is only one. Thus, while defendants may argue for admittance, counsel expertise and case-building energy is likely focused in multiple directions. This has the effect of minimizing the strength of any single defense argument for admittance of polygraph evidence. Compounding the calculation is the fact that for the defendant this is a one-shot game. Contrast this with law enforcement in the context of DNA analysis. Law enforcement is a repeat player in judicial proceedings. Thus, the incentive to obtain admissibility of DNA evidence is greater so as to prosecute with DNA evidence in future rounds of play. Defendants, on the other hand, are focused only on this particular case. They may feel they can win it without polygraph evidence, so they take the risk of nonacceptance rather than mounting a full-scale scientific assault on the court. This creates a "weak" record. The prosecution, on the other hand, will fight for a long-term payoff when the admissibility of evidence is crucial to this case as well as many more cases in the foreseeable future—as it was with DNA. Similarly, when the payoff is small for courtroom admission, the calculus does not encourage law enforcement to allocate resources to seeking admissibility, especially when polygraphy has done the heavy lifting prior to the trial. Thus, polygraph policy advocates have different incentive structures than DNA advocates. Because of this, we may expect to find different appellant strategies and less expert testimony.

Another problem with polygraph evidence is that it is an indirect measure of a state of mind. By contrast, DNA is a direct measure of matching nucleotide sequences or matching variable tandem repeats. Logically speaking, polygraph evidence is much more vulnerable to scientific attacks. There are more steps in the process and more potential points of disagreement. For judges trying to determine the validity and reliability of the science, polygraph evidence can quickly appear unreliable and unacceptable to the relevant scientific community. DNA has two weak points: the human comparison of DNA samples and the crafting of statistical likelihood estimates. The human comparison of DNA samples is a one-step process—the forensic scientist examines the slides to see if the bands appear to match. By contrast, polygraphy has several weak points: the theoretical link between physiological symptoms and emotional state and the cause of that emotional state, the human crafting of the interview process, and the human comparison of that process. Human behavior and conjecture are present and problematic for polygraph evidence. Thus, logically, this particular "way of knowing" will be subject to more doubt. Or, to put it another way, DNA is a less contentious science than polygraphy. Therefore, we would expect much more contention with the latter, and we would expect courts to be reluctant to approve of the science for legal purposes.

We may also expect several degrees of acceptance and compromise measures to emerge in the judicial process. The wiggle room created by the dubious science creates wiggle room for judicial outcomes. Where DNA was simply admissible or inadmissible, polygraphs might be partly admissible or sometimes inadmissible. For example, it is possible for the parties to agree in advance to admit the evidence. In that case, polygraph information would become stipulated evidence. What can the courts say to this? Will they accept the stipulation or will they reach beyond it to examine the validity and reliability of the evidence? Another example is the admission of the evidence for some criminal proceedings and not others. In that case, jurisdictions may rule to allow polygraph evidence for extra-trial proceedings (pre-sentencing hearings, post-sentencing hearings, etc.) while disallowing it in the trial court.

In terms of political maneuvering, we would expect proponents and opponents to exploit this uncertainty to their advantage. The state's power is greatest before the trial or during initial questioning of a defendant. Thus, the state could seek a stipulation agreement early on in the criminal process. Many times defendants will offer to take polygraphs fairly soon in the process, hoping to convince law enforcement of their innocence. At the same time, the state will ask the defendant to sign a stipulation agreement. The state then has what it needs to be able to present polygraph evidence at a trial without proving validity or reliability of polygraphs. Courts admitting stipulated polygraph evidence may still require expert testimony to assist a jury and allow for cross-examination of experts to weaken the effect of the results. Prosecutors can generally weather this extra burden fairly well, having several polygraph experts readily at hand.

The defendant's power is greatest in extra-trial proceedings when evidentiary standards are relaxed in the name of due process fundamental fairness. Especially in capital sentencing cases, the scales are tipped in the defendant's favor to present as much mitigating evidence as possible: "When important decisions in industry and government are made with the help of polygraph tests, it seems strange to refuse any use of such information to determine whether to impose a life or death sentence. To perpetuate such a ban is to say that the leaders of government, law enforcement, and industry are all wrong in deciding what to consider in making important decisions. But polygraph techniques have improved to the point that we cannot realistically make that claim" (Concurring opinion, *State v. Harrod* 26 P.3d 492 [Ariz. 2001]). Other cases made a similar observation: polygraphs are routine in other areas of government, including as a condition of probation (see *State v. Lumley*, 977 P.2d 914 [Kan. 1999]). Defendants often seek to include favorable polygraph evidence at these stages of the criminal process, arguing that results are at least reliable enough for official uses.[4]

The legal literature on polygraph evidence does not always recognize gatekeeping as a political process, where actors are pursuing their interests within the parameters of the environment. For this reason, legal scholars can have trouble explaining the outcomes from a systemic viewpoint. As with DNA, the following analysis looks at the role of institutional/organizational variables, attitudinal variables, and legal variables in gatekeeping outcomes. It is my contention that judicial outcomes are the result of a less than robust science interacting with the political process.

The Gatekeepers and Polygraph Evidence

When faced with the question of admitting polygraph evidence as scientific evidence, courts had multiple possible responses. These responses can be placed roughly along a spectrum. On the side of rejection were decisions with per se findings of inadmissibility. These decisions maintained polygraphs to be inadmissible as a matter of general policy. There were also rejection decisions that explicitly forbid admittance by stipulation. In terms of acceptance, there were decisions admitting stipulated results and admitting results in extra-trial proceedings. There were also decisions admitting unstipulated results as a matter of judicial discretion. This section will examine the distribution of gatekeeping outcomes and jurisdictional patterns.

General Findings

The analysis reported 165 cases from 1933 to 2001 in which state supreme courts ruled on the admissibility of polygraph evidence as a question of scientific validity and reliability.[5] Each case was coded for the holding with regard to polygraph evidence. The following descriptive statistics break this down by year, state, and holding.

In table 4.1, the initial pattern of acceptance and rejection is shown across time. State abbreviations are arrayed across a timeline to indicate a state supreme court polygraph gatekeeping decision. Policy decisions admitting polygraphs are separated from decisions rejecting polygraph evidence. Several observations are immediately forthcoming from table 4.1. First, the majority of decisions (136) have rejected polygraph evidence as valid or reliable for criminal judicial proceedings. These decisions have a per se rule against admittance of results, even stipulated results.[6] The other twenty-nine decisions are cases where the court agreed to admit polygraph if the parties stipulated (e.g., agreed to ahead of time) to admission or if it was a matter of judicial discretion.

Second, the timeline for polygraph evidence is significantly longer than the DNA timeline. State supreme court decisions range from 1933 to the present, whereas DNA has only been on the judicial radar since 1989. This timeline also shows a pattern of complete rejection until 1960, with rejections dominating until 1974. This was not the case with DNA. From the beginning DNA had a mixed record in state supreme courts. As with DNA, the amount of state activity over time is not steady for polygraph evidence. The timeline shows sporadic cases until 1955 when the first large wave of cases arrived. Ten jurisdictions ruled on polygraph evidence from 1933 to 1954. Seventeen more jurisdictions ruled on polygraph evidence from 1955 to 1966, and a total of twenty-five decisions were made during this period. The timeline shows that the majority of decisions were made in the next twenty years. From 1968 to 1988, roughly half (eighty-three) of

TABLE 4.1

ADMISSIBILITY OF POLYGRAPH EVIDENCE IN STATE SUPREME COURTS
(SCIENTIFIC RELIABILITY)

Admissible		Inadmissible
	1933	WI
	1940	
	1941	
	1942	MI
	1943	
	1944	
	1945	MO
	1946	
	1947	KS
	1948	
	1949	NE
	1950	ND
	1951	OK, TX
	1952	FL, MN
	1953	
	1954	ME
	1955	MI, MS
	1956	PA
	1957	CA, KY, MO
	1958	TN, VA
	1959	SC
IA	1960	MT
	1961	MS, NM, NC
	1962	HI, ME, MN, NJ, AZ
	1963	IL, MA
	1964	KY
	1965	NH
	1966	CA
	1967	
MO	1968	AK, MI, OR
	1969	NY
	1970	AK, WA, IA

(*continued*)

TABLE 4.1

ADMISSIBILITY OF POLYGRAPH EVIDENCE IN STATE SUPREME COURTS
(SCIENTIFIC RELIABILITY) (CONTINUED)

Admissible		Inadmissible
	1971	RI
NJ	1972	IN, LA, SD
	1973	LA, PA, SC, TX
WI, UT, NM, MA	1974	NV, ND, WA, OK
NM, FL	1975	CT, GA, OK, OR, NC
	1976	GA, LA, NE, NH, PA, KS, SD
GA, WY	1977	MI, AZ, AR
OH, NV	1978	DE, MT, TX, MT, ND, MA, KS
WI, NC, IA	1979	NH, WV, UT
	1980	AL, MD, MO, MT, UT, DE
	1981	CO, IL, WI
WA	1982	MA, WA
	1983	IL, NC, PA, FL
UT	1984	LA, OR
NM	1985	MN, SD, SD, VT
ID	1986	VA, NV, IN
	1987	ND, OR, UT
	1988	RI
	1989	AL, CA, MA, MT, ID
SD, OH	1990	NV
	1991	KY, MT
	1992	
	1993	
ID, NM	1994	
	1995	MS, WV, CT
IN	1996	CA, NY, RI, UT, AZ
	1997	CT, NE, UT
	1998	ND, MS
KS, ID	1999	CO, SC, KS
NM	2000	LA, NC, NV, KS
	2001	TN, IN, AZ
	2002	
Total cases 29		136

all state supreme court decisions on the scientific admissibility of polygraph evidence were decided. This time period is also notable for the surge in decisions admitting polygraph evidence. During this period numerous third-party reports surfaced, especially widely cited conflicting reports by the U.S. Department of Defense (1984) (highly accurate) and by the U.S. Congress Office of Technology Assessment (1983) (significant error rates). The late 1990s experienced a modest surge in polygraph case activity, especially after the ambiguous conclusions of the U.S. Supreme Court in *United States v. Scheffer*, 523 U.S. 203 (1998), the only opinion issued by the court since 1923 on the subject of polygraph reliability.[7]

Like all gatekeeping decisions, state supreme court polygraph cases either rejected or accepted polygraph evidence. Yet, unlike the strictly dichotomous approach to DNA evidence, polygraph cases can also be arrayed in a continuum. This is because state supreme courts found themselves dealing with a mutation in admissibility law. In judicial proceedings, parties can agree ahead of time not to contest the introduction of certain evidence in a stipulation agreement. As discussed above, stipulation agreements became a routine part of polygraph interviews conducted by police. Thus, while courts were deciding whether to admit polygraph evidence, they also had to decide what to do when the parties had agreed ahead of time to admit the results of the polygraph test. Courts essentially had two options with stipulated results: they could reject them anyway, or they could allow them to be admitted. This has the effect of generating a multiplicity of judicial responses.

In table 4.2, this variety of polygraph holdings is listed across time. Courts displayed several responses when deciding whether to admit polygraph evidence. As this table shows, there is a continuum of court response, from outright per se rejection to liberal admission under judicial discretion. As this table illustrates, these holdings fall roughly in a bell shape: 4 percent are per se rejection, 10 percent rejected even stipulated results, 51 percent simply ruled polygraph evidence inadmissible, 18 percent would have admitted stipulated results, 12 percent admitted stipulated results, and 5 percent admitted the evidence as a matter of judicial discretion. Furthermore, as matter of current policy, thirty-four jurisdictions do not admit polygraph evidence, even with a stipulation. Sixteen jurisdictions admit polygraph evidence with stipulations. In addition, four states rendered decisions allowing extra-trial consideration of polygraph evidence (Idaho, Louisiana, Kansas, and Michigan), and two states allow judicial discretion for non-stipulated results (Idaho and New Mexico).

DNA and polygraph have significantly different acceptance patterns. When the mean judicial holding is charted over time, these differences become stark. With polygraph, the mean holding begins with a stable jurisprudence of rejection. The mean holding edges toward admissibility beginning in the early 1970s when some courts began liberalizing their polygraph jurisprudence. In the mid-1980s the pendulum begins to swing back to a mean holding of inadmissibility. In general, the overall polygraph pattern is divergent and shows jurisprudence

TABLE 4.2

HOLDINGS ON ADMISSIBILITY OF POLYGRAPH EVIDENCE IN STATE SUPREME COURTS (DECISIONS BASED ON SCIENTIFIC RELIABILITY/VALIDITY)

Year	Inadmissible Per Se	Inadmissible Even w/Stipulation	Inadmissible in General	Inadmissible Need Stipulation	Admissible Have Stipulation	Admissible Jud. Discretion
1933			WI			
1942			MI			
1945			MO			
1947			KS			
1949			NE			
1950			ND			
1951			OK, TX			
1952			FL, MN			
1954			ME			
1955			MI, MS			
1956			PA			
1957			MO, CA, KY			
1958			TN, VA			
1959			SC			
1960			MT			IA
1961			NM, MS	NC		
1962			HI, ME, MN, NJ	AZ		
1963			IL, MA			
1964			KY			
1965			NH			
1966			CA			
1968			AK, MI, OR		MO	
1969			NY			
1970		AK	WA	IA		
1971			RI			
1972			SD, LA, IN		NJ	
1973		TX, LA	SC, PA			
1974			NV, ND, WA	OK	WI, UT, MN, MA	
1975	CT	OK	OR, GA	NC	FL	NM

(*continued*)

TABLE 4.2

HOLDINGS ON ADMISSIBILITY OF POLYGRAPH EVIDENCE IN STATE SUPREME COURTS
(DECISIONS BASED ON SCIENTIFIC RELIABILITY/VALIDITY) (CONTINUED)

Year	Inadmissible Per Se	Inadmissible Even w/Stipulation	Inadmissible in General	Inadmissible Need Stipulation	Admissible Have Stipulation	Admissible Jud. Discretion
1976		GA	LA, NE, NH, PA	KS, SD		
1977			MI	AR, AZ, MA, KS	WY, GA, OH, NV	
1978			MT, MT, ND, TX, DE			
1979		WV	NH	UT	WI, NC	LA
1980		MO	MT, MD, AL	UT, DE		
1981	CO	IL, WI				
1982				MA, WA	WA	
1983		NC, PA	IL	FL		
1984			OR, LA		UT	
1985			MN, SD, SD, VT			NM
1986			VA	IN, NV		ID
1987		OR	ND	UT		
1988			RI			
1989	AL	MA	CA, MT	ID		
1990				NV	SD	OH
1991		KY	MT			
1994						NM, ID
1995	CT,	WV	MS			
1996			CA, RI, NY	UT, AZ	IN	
1997	CT		NE	UT		
1998			ND, MS			
1999	CO		SC	KS	KS	ID
2000		NC	LA	NV, KS		NM
2001			TN	IN, AZ		
TOTAL	6 (4%)	16 (10%)	85 (51%)	30 (18%)	19 (12%)	9 (5%)

Six states (FL, MA, MO, NC, OK, WI) admitted stipulated evidence at some point during their jurisprudence. Five states admit evidence extra-trial (ID, LA, KS, MI) or at the trial court's discretion (ID, NM) in addition to allowing stipulated results.

destabilizing after 1968. This stands in direct contrast to the pattern of DNA jurisprudence. With DNA, the mean holding was initially destabilized, with holdings all over the map. The mean holding dipped from 1991 to 1994, the period where judicial contention is most evident, to bottom out at roughly halfway between acceptance and rejection in 1993. After 1997, the system clearly edges toward admissibility. In general, the DNA pattern is convergent and shows jurisprudence stabilizing in favor of acceptance after 1997.

States varied in the number of polygraph decisions undertaken during this period. When the number of decisions is broken down by state, five states had one high court decision on the validity and reliability of polygraph evidence in criminal trials. The average number of decisions was approximately three decisions per jurisdiction. The mode number of decisions was two or four (thirteen states each).

These general findings create the puzzles investigated later in the analysis: Why did some states admit while other states rejected? Why were some states more active than others? What is significant about the noticeable increase in acceptance after 1974? Before we can begin the quest for causation, it is necessary to take a closer look at the nature of the holdings with regard to polygraphy.

Black Robes in Action

Interesting patterns of jurisdictional behavior emerge from the pattern of polygraph admissibility. First, there is a distinct group of states whose initial ruling on polygraph evidence has not been meaningfully revisited. These *single-decision* jurisdictions only had one decision on the reliability and admissibility of polygraph evidence. Those decisions mostly favored rejecting polygraph evidence (Hawaii, 1962; Maryland, 1980; Vermont, 1985). Two single-decision jurisdictions, however, admit polygraph evidence with a stipulation (Arkansas, 1977; Wyoming, 1997). The presence of early decisions implies that some jurisdictions did not feel the need to revisit initial precedents with each new wave of polygraph literature. It also suggests this single-decision phenomenon was not limited to latecomers who simply wade into the calm waters of a receding controversy. Indeed, Arkansas took a controversial position only three years after the first stipulation decision (Wisconsin, 1974). Wyoming remained aloof the longest, finally deciding to admit upon stipulation. As can be seen in table 4.1, the late 1990s represent a small resurgence in polygraph decisions coinciding with the above-mentioned U.S. Supreme Court statement of continued controversy.

Second, there are two groups of states in which initial decisions and subsequent decisions varied (see table 4.3). These groups of *dynamic* jurisdictions are distinguished by the fact that their jurisprudence changed in the course of the polygraph debate. The first group moved from rejection of polygraph evidence to acceptance of stipulated results. The second group moved from rejection to experimentation with stipulated admission to rejection again. These dynamic jurisdictions are evidence of the way jurisdictions wrestled with the question of

TABLE 4.3

JURISDICTION PATTERNS FOR DYNAMIC JURISDICTIONS

State	(Rejection Decisions)	Decisions Admitting w/Stipulation	(Rejection Decisions)
Moving from rejection to acceptance of stipulated results			
Delaware	(1978)	1980	
Georgia	(1975, 1976)	1977	
Indiana	(1972)	1986, 1996, 2001	
Kansas	(1947)	1976, 1978, 1999, 2000	
Nevada	(1974)	1978, 1986, 1990, 2000	
New Jersey	(1962)	1972	
New Mexico	(1961)	1974, 1975, 1985, 1994, 2000	
South Dakota	(1972)	1976[†]	
Moving from rejection to acceptance of stipulated results to rejection of stipulated results			
Florida	(1952)	1975	(1983)
Massachusetts	(1963)	1974, 1978, 1982	(1989)
Missouri	(1945, 1957)	1968	(1980)
North Carolina		1961, 1975, 1979	(1983)
Oklahoma	(1951)	1974	(1975)
Wisconsin	(1933)	1974, 1979	(1981)

This analysis does not include extra-trial rulings.

[†]South Dakota was an unusual jurisdiction. In 1972 it ruled against polygraph evidence. In 1976 it admitted stipulated results. In 1985 it twice rejected non-stipulated admittance. In 1990 it reiterated its rule to admit stipulated results. Utah was an unusual jurisdiction as well. In 1974, 1979, 1980, 1984, and 1987 it continually held to a rule of admitting only stipulated results. In 1996 and 1997 it ruled against unstipulated results. Washington followed this pattern, rejecting polygraphs in 1970 and 1974, accepting stipulated results in 1982, and rejecting unstipulated results again in 1982.

polygraph admissibility. The years of turmoil appear to be the 1970s and the early 1980s. Looking only at the second group, it is clear to see that the 1970s was the time for experimentation; the 1980s was the decade for returning to a rejection rule. Early rejection prior to 1972 was the norm for both groups.

Third, there was a group of states with multiple decisions in the same direction (see table 4.4). This group of *monotonic* jurisdictions is characterized by consistent

TABLE 4.4

JURISDICTION PATTERNS FOR MONOTONIC JURISDICTIONS

State	Year
Monotonic Decisions Admitting Polygraph	
Arizona	1962, 1977, 1996, 2001
Idaho	1989, 1986, 1999
Iowa	1960, 1976
Ohio	1978, 1990
Monotonic Decisions Rejecting Polygraph	
Alaska	1980, 1989
California	1957, 1966, 1989, 1996
Colorado	1981, 1991*
Connecticut	1975, 1997*
Illinois	1963, 1981, 1983
Kansas	1957, 1964, 1991
Louisiana	1972, 1973, 1976, 1984, 2000†
Maine	1954, 1962
Minnesota	1952, 1962, 1985
Mississippi	1955, 1961, 1995, 1998
Montana	1960, 1978, 1980, 1989, 1991
Nebraska	1949, 1976, 1997
New Hampshire	1965, 1976, 1979
New York	1969, 1996
North Dakota	1950, 1974, 1987, 1998
Oregon	1968, 1975, 1984, 1987
Pennsylvania	1959, 1973, 1976, 1983
Rhode Island	1971, 1988, 1996
South Carolina	1959, 1973, 1999
Tennessee	1958, 2000
Texas	1951, 1973, 1978, 1998
Virginia	1958, 1986
West Virginia	1979, 1995

This analysis excludes extra-trial decisions.
*Colorado and Connecticut were the states with the strongest statements of rejection of polygraph evidence.
†Louisiana did allow extra-trial admittance in a 1979 decision.

jurisprudence in divergent directions, with some states steadily admitting polygraph evidence while most states consistently reject polygraph evidence. Multiple visits to the same question, without any shift in established policy, may seem insignificant. These were genuine attempts by the appellants to seek change, however, especially in the "rejecting" jurisdictions. The following quotes give the flavor of the majority of these cases:

> On a proper showing, defendants must, from time to time, be permitted to demonstrate that advancement in scientific acceptance has occurred. [Here], defendant offered only to call the polygraph examiner to establish the manner in which the test was conducted, an offer that was not sufficient to establish the admissibility of the results of the test or the examiner's opinion regarding defendant veracity. Absent an offer of proof that the polygraph is now accepted in the scientific community as a reliable technique, the evidence was presumptively unreliable and inadmissible. (*People v. Harris*, 767 P.2d 619 [Cal. 1989])
>
> To create a record sufficient to allow this court to consider altering long-standing Connecticut law barring polygraph evidence, the defendant bore the burden of demonstrating that traditional reasons for the rule were no longer applicable. Even under the *Daubert* rule, evidence must be shown to be relevant and reliable to be admissible. Here, defendant bore the burden of creating a factual record before the trial court that the polygraph test possesses sufficient reliability to justify its introduction as scientific evidence. The defendant's only evidence arguably relevant to his claim of the increased reliability of polygraph testing evidence was his offer of proof that the Hamden Police investigating the murder asked several individuals to agree to take a test.... Evidence that the police made limited use of the test in the course of a particular murder investigation does not constitute sufficient proof of reliability to justify overturning the traditional rule. (*State v. Esposito*, 670 A.2d 301 [Conn. 1996])

As these quotes demonstrate, repeated attempts to change precedent often lacked sufficient expert evidence. Yet what does this tell us about *dynamic* jurisdictions? Do those decisions represent cases where sufficient arguments were mounted? Not exactly. The majority of the stipulation cases followed the lead of a single jurisdiction: Wisconsin. In the 1971 *State v. Stanislawski*, 216 N.W.2d 8, case, the Wisconsin Supreme Court created a four-step jurisprudence for judicial acceptance of stipulated polygraph evidence. The bar was still significantly high, requiring the parties to present expert evidence at a pretrial hearing. Several subsequent cases in other jurisdictions adopted the *Stanislawski* requirements. When Wisconsin overturned the *Stanislawski* decision in 1981, this similarly precipitated changes in several other jurisdictions. Thus, while some dynamic jurisdictions at least conceded the advancement of polygraphs, much more of the movement had to do with the viability of a middle-ground acceptance of stipulated results.

Disagreements on the Bench

Another feature of the judicial processing of polygraph information is the degree of debate present in a given case. Historically, judicial debate in the form of concurring and dissenting opinions has been rare on state supreme courts (Jaros and Canon 1971). Because debate is rare, much could be gleaned from a closer examination of the character of debate among colleagues faced with the same facts in the same jurisdiction.

Nineteen (12 percent) decisions in fourteen jurisdictions contained concurring or dissenting opinions where judges disagreed with the majority ruling on polygraph admissibility (see table 4.5). The character of the debate is explained with annotations of the majority opinion and corresponding concurring or dissenting opinions. Twelve of the judicial debates occurred in the eight years between 1974 and 1982—echoing the debate among courts in the larger judicial context. Almost all the debates are about the prudence of allowing or foreclosing stipulated results. This period is characterized by a destabilizing judicial context, and individual courts mirror this destabilization with the increase of judicial debate on their own bench. The development of a polygraph "middle ground" (admitting stipulated results) destabilized the consensus against polygraph admittance. Prior to this time, courts predominantly had two choices: to accept or reject polygraph evidence. This middle-ground mutation in the debate destabilized the legal environment, creating an increase in the variation of judicial outcomes.

Several arguments were advanced by dissenting and concurring opinions. Individual judges felt compelled to speak out against continued rejection or against the idea of admitting stipulated results. They evidently drew their comments from ideas at work in the larger legal community. Individual judges also tried to make connections between the reliability of polygraph evidence and the reliability of other accepted evidence, such as eyewitness testimony or psychiatric evaluations. Court majorities answered these reliability comparisons by distinguishing between lay or professional opinions and the undue weight juries might give to the "scientific" nature and perceived objectivity of polygraph evidence. Other individual judges pointed out inconsistencies in judicial logic, especially the fact that stipulations do not increase the reliability of polygraph evidence or the fact that other institutions (law enforcement, industry, and the military) consider the evidence reliable for their purposes.

Polygraph jurisprudence is clearly one area where "reasonable people" can disagree. It is not surprising to find intra-court disagreements echoing inter-court inconsistencies.

Changing Analysis over Time

Another preliminary consideration involves the way courts go about reaching their conclusions. It may be that different methods lead to different conclusions, thus partly explaining the variance in the outcomes. Judicial decisions were coded for

TABLE 4.5

SUMMARY OF JUDICIAL DEBATE

	State	Court Majority	Concurring or Dissenting Argument
1974	Massachusetts	Admit with stipulation	(Three dissenters) Wait for more info, appoint a commission
1974	Oklahoma	Admit with stipulation	(Concurrence) Still not scientifically accepted
			(Dissent) Should reevaluate when there is a competent witness, such as a member of the Amer. Polygraphers Assoc.
1976	Georgia	Inadmissible	Advanced enough to allow stipulated admissibility
1976	Louisiana	Inadmissible	Should admit or at least allow more discussion
1976	Nebraska	Inadmissible	There maybe some circumstances allowing admission
1977	Georgia	Admit with stipulation	Stipulation does not make results reliable
1978	Texas	Inadmissible	Should admit, no less reliable than psychiatry
1979	Louisiana	Inadmissible per se	(Dissent) would not adopt a per se rule
		Extra-trial admissible	(Concurrence) would not admit extra-trial either
1979	Wisconsin	Admit with stipulation	Should overrule *Stanislawski* (admitted stipulated results)
1980	Missouri	Inadmissible per se	Should admit with stipulation
1980	Montana	Inadmissible	State's idea to take the test, should have been admitted
1981	Colorado	Inadmissible per se	Should allow stipulated results
1982	Washington	Admit with stipulation	Stipulations do not increase reliability
1989	Idaho	Requires stipulation	Should at least be admissible for fundamental fairness
1989	Massachusetts	Inadmissible per se	Should admit under limited circumstances
1997	Connecticut	Inadmissible per se	Should at least allow a hearing on admissibility at trial
1997	Utah	Inadmissible	We should revisit polygraph holdings
1999	Idaho	Judicial discretion	Not yet reliable enough
2001	Arizona	Admit with stipulation	Polygraph jurisprudence inconsistent with law enforcement

the type of analysis undertaken by the court. These fell into two main categories: (1) analysis that cited precedent or general rules (e.g., *American Law Review* citations) and (2) analysis that attempted an independent review of the relevant literature and expert commentary. Precedent or general rules were used for primary analysis in 58 percent of cases. Relevant literature and expert commentary was used for primary analysis in 31 percent of cases. Eleven percent of cases used some other form of analysis. Court analysis was somewhat related to the year of the decision (ANOVA $F = 1.45$, $p = .052$), suggesting the increase of analyses of relevant literature and expert testimony. When the mean court analysis is charted over time, there is a pattern of increased use of relevant literature beginning in the 1960s.

Logically, this shift in analysis is to be expected. Precedent from the original polygraph case (*Frye v. United States*, 293 F. 1013 [1923]) and general rules stated thereafter were the dominant cues for decision making prior to the early 1960s. There were few polygraph experts to generate expert commentary during this time. By 1963 and later, commentary, professionalization, and literature increased, both from polygraphers and legal analysts. This meant that courts had a wider variety of sources from which to draw their conclusions. This shift in analysis also changed the direction of many cases. There is a clear pattern between analysis and outcome. Court analysis was significantly related to judicial outcomes in a statistical comparison (phi $= .525$, $p = .000$). In terms of acceptance patterns, courts examining relevant literature and expert testimony, as well as those using some other type of analysis, were more likely to admit polygraph evidence than courts adhering to precedent or a general rule.

Taken together, the behavior of jurisdictions, the increase in judicial debate, and the different patterns of analysis suggest strong evidence for an overall pattern of divergence with regard to polygraph evidence. This pattern stands in stark contrast to the convergence experienced by DNA evidence. This does not mean, however, that the same clues to judicial behavior are not at work. There may be important legal, attitudinal, and political factors associated with these outcomes, which explain state polygraph variation in a manner consistent with DNA variation. The following analysis will examine the role of legal standards, partisanship, region, and several political leverage variables for a significant pattern.

Explaining Polygraph Gatekeeping: Legal Considerations

One clue to judicial behavior suggests that gatekeepers are operating according to some standard of who (and what) can (or cannot) be admitted. If the rules of admissibility vary from jurisdiction to jurisdiction, this would presumably affect the chances of admittance. In chapter 2 this was conceptualized as a "threshold" at work in a jurisdiction. Jurisdictions with lower thresholds (or lower legal standards) will presumably experience more acceptance than jurisdictions with higher thresholds.

By way of review, there are essentially three standards used in the judicial admittance of scientific evidence: the relevancy standard (state versions of the

TABLE 4.6

COMPARING LEGAL STANDARDS AND JUDICIAL OUTCOMES
FOR POLYGRAPH EVIDENCE

Legal Standard in the Decision	Cases	Decisions Admitting	Decisions Not Admitting
Frye or "general acceptance"	151	21	130
Daubert (one of four criteria)	7	1	6
FRE 702	7	6	1

Likelihood Ratio 20.998 (Sig. .000)
Kendall's Tau-b .224 (p = .077)

Federal Rules of Evidence Section 702, hereinafter FRE 702), the general acceptance standard (*Frye* standard), and the *Daubert* standard (four potential criteria for a judicial decision based on the *Daubert* case). In terms of polygraph evidence, the legal standard at work in a jurisdiction is expected to influence judicial outcomes. The more liberal the standard, the more likely it is the jurisdiction will find the science acceptable. The more difficult the standard, the more likely scientific evidence will be rejected. Even the strictest standard, the *Frye* "general acceptance," however, has an important built-in dynamic: time. The diffusion of a novel scientific technique may result in growing acceptance by the scientific community. So, even in *Frye* districts, we may expect to see a pattern where *Frye* districts that initially rejected polygraphy later found it to be "generally acceptable" simply because of widespread use and acceptance by the scientific community over time. It is also possible that "liberal" standards provide science with a legal foothold that eventually allows it to overcome more conservative standards. For instance, as some jurisdictions admit polygraphy, the use and acceptance of the technique grows, possibly resulting in later admittance by more reluctant jurisdictions.

There was little variance in the judicial standard used by polygraph courts (see table 4.6). Almost all of the decisions (91 percent) occurred under a *Frye* general acceptance analysis. This is due to the fact that most decisions occurred before the *Daubert* decision was issued in 1993. In the *Frye* cases, only 20 out of 150 decisions admitted polygraph evidence. In the cases where the FRE 702 standard was used, 6 out of 7 cases admitted the evidence. This does suggest that the use of a different standard could have a different outcome, yet, due to the limited comparison data, the relationship was not statistically significant.

A closer look at the cases (table 4.7) shows the six FRE 702 cases also had the most liberal holdings for polygraph admittance: judicial discretion. Those cases were more likely to leave polygraph admissibility to individual judicial discretion. Table 4.7 also shows the *Daubert* standard was also the least liberal; six of seven *Daubert* holdings were inadmissible. For the *Frye* cases, one hundred

TABLE 4.7

COMPARING LEGAL STANDARDS AND JUDICIAL HOLDINGS FOR POLYGRAPH EVIDENCE

Legal Standard	Inadmissible Per Se	Inadmissible Even w/Stip.	Inadmissible in General	Inadmissible Need Stip.	Admissible Have Stip.	Admissible Jud. Discretion
Frye	4	15	81	30	19	2
Daubert	2	1	3	0	0	1
FRE 702	0	0	1	0	0	6

decisions held polygraph evidence to be inadmissible, and an additional twenty-nine ruled that stipulations were needed to admit polygraph evidence. In general, even with little variance in legal standards, a pattern has emerged for polygraph evidence. While not as convincing as the patterns found in DNA analysis, these patterns do not contradict those findings either. Liberal standards are associated with liberal outcomes and liberal holdings. Strict standards are likewise associated with conservative outcomes and conservative holdings.

The Role of Legal Options

While there is little variation for a comparison of legal standards and judicial outcomes, part of the theoretical concern can still be investigated empirically. The theoretical concern is with the role of law in judicial outcomes. This has been defined as the legal standard at work in the jurisdiction. It can also be defined as the legal context. Polygraph evidence was staunchly rejected on scientific grounds, even while law enforcement and others continued to find useful applications. This reality was part of the pressure for acceptance. The development of the stipulation and the middle-ground approach *changed* the legal context—not in terms of calculus, but in terms of judicial options. The legal options provided by the legal environment inform the calculus of individual decision makers. In other words, the development of the stipulation option changed the way courts could approach their gatekeeping policy. Here, with polygraph evidence, we see the addition of a third option destabilizing and complicating the debate—with the obvious effect of allowing polygraph evidence and testimony into the courtroom. This had important implications for polygraph jurisprudence and the judicial view of the science. Many courts referred to the admission of stipulated results as a policy "experiment" or as a "crack in the door"—ideas recognizing that the scientific debate is unresolved. Courts lamented that there is no agreement on how "reliable" a science must be to be admitted to court, and certain courts were explicitly willing to allow for limited admissibility as a way to open the system to polygraph applications in the judicial process.

From a theoretical viewpoint, the popularity of stipulated admissibility can be explained as minimizing cognitive work while maximizing psychological satisfaction. Allowing stipulated results greatly decreased the need for the use of complicated calculi, such as determining whether polygraphy was generally accepted. It also increased psychological satisfaction because judges felt sufficient safeguards were in place and because now the science could at least find some entry into the judicial process. Those in the system were obviously dissatisfied with the earlier choice between two less-than-satisfactory solutions—either rule that polygraph evidence is completely reliable or rule that it can never be used in case-by-case analysis. Stipulated admissibility opened the door for the evolution of jurisprudence, especially to the individual judges who were not comfortable foreclosing any use of polygraph information at trial. It was also more psychologically satisfying because it created a way for the system to absorb advances in the science and technology of polygraph evidence: as it became more reliable, stipulations might increase, thus providing for direct evaluation by affected parties—a solution that clearly found favor with many judges.

From a theoretical standpoint, the backlash against stipulated results can also be explained. Among those individuals who defined psychological satisfaction as stable, simplified calculi, stipulated admissibility provided no comfort. These judges felt more comfortable with a black-and-white approach where science defined as controversial is science with no place in judicial proceedings. This reduced the cognitive work of these judges by simplifying their calculus. It also was more psychologically satisfying for those who wanted to make sure the legal system was using only the best, most reliable information for decision making.

Legal variables certainly had a role to play in polygraph jurisprudence—whether it was the lowering of a threshold or the development of a policy mutation, such as stipulated admissibility. The development of a middle ground rule and its interaction with the legal standards clearly accounts for a portion of the pattern of polygraph acceptance and rejection.

Explaining Polygraph Gatekeeping: Democrats, Republicans, and the West

A second clue to judicial behavior suggests that gatekeepers are operating according to their own attitudes toward the science and the politics it represents. The simple expectation is that certain gatekeepers will behave differently because of their political preferences. If gatekeepers prefer a science and/or support its advocates, they will be more likely to rule in favor of admissibility. Conversely, another set of gatekeepers might reject a science as a way to reject the political goals of those advocating for its admissibility. This clue suggests that if preferences vary from jurisdiction to jurisdiction, admissibility might vary as well. In chapter 2 this was conceptualized as an "attitudinal support." Perhaps the most intuitive example is the example of judicial attitudes familiar to most trial lawyers. Trial lawyers know that some judges are more lenient (or more severe) toward certain litigants. This idea of a friendly (or hostile) court could easily carry over to scientific admissibility, especially when admissibility is coupled with either the defendant or the prosecution advocating a position. There is also a more complex conceptualization of attitudinal support. The complex picture supposes that yes, gatekeepers may have their own attitudes, but they are operating in a political context, such as a particular region of the country, which may constrain those attitudes. For this reason, we cannot be too hasty to assume a direct relationship between judicial preference and gatekeeping outcome. If attitudinal correlations exist, however, the preferences of someone (gatekeeper or community) might be work.

Partisanship

As with the DNA cases, it is possible to look for an attitudinal explanation by comparing partisanship and policy making. To test this hypothesis, it was first necessary to obtain partisan information on the courts and judges. This was accomplished through a two-step process. First, the names of the judges serving on the cases were obtained. For most courts, these names were found on the opinions. Second, most judges were labeled with a partisan label (Democrat or Republican). These labels were obtained either by a direct label (in the *Judicial Yellowbook*) or an indirect label extrapolated from the party of an appointing governor.

There were 485 named judges on the fifty state supreme courts spanning seventy years of polygraph jurisprudence. This yielded a total of 634 judicial votes on polygraph evidence. For most of these judges, however, it was not possible to supply a partisan label. For many early cases (pre-1965) it was difficult to obtain participating judges and judicial partisanship. Several early opinions feature only the name of the opinion writer. Available state-level judicial directories were commercially published after 1970. Where it was possible to label judges, 146 polygraph votes were cast by Democratic judges and 68 were cast by Republican judges. An analysis of this subgroup of identifiably partisan judges might provide some support for attitudinal differences.

The first cut of the analysis examined the direction of partisan judges who disagreed with their colleagues on the bench. The examination of dissent patterns yielded a detectable partisan pattern. Ten Democratic judges joined in polygraph admissibility dissents. Of these ten, only two dissents argued for rejection of polygraph evidence. The other eight argued for admissibility, two for stipulated results and six for admitting with trial court discretion. This implies that these Democrats were more likely to support admitting polygraph evidence. By contrast, only one dissent was joined by an identifiable Republican. This dissent argued to reject polygraph evidence. Another nineteen unidentified judges dissented, eight against admissibility, one in support of stipulated admissibility, and ten for admittance under judicial discretion. Clearly, the identifiable Democrats overwhelmingly supported polygraph evidence.

TABLE 4.8
PARTISANSHIP AND JUDICIAL VOTES TO ADMIT POLYGRAPH EVIDENCE

	Inadmissible	Admit w/ Stipulation	Admit
Democrats	40% (59)	43% (63)	16% (24)
Republicans	45% (31)	47% (32)	7% (5)

Chi-square .510 (p=.475)
Likelihood Ratio .320 (p=.572)
Phi .049 (p=.475)
One-way ANOVA F .507 (p=.477)
Non-labeled judges are largely from before 1975, when partisanship was least likely to be measured. Over half the decisions occurred before 1975.

TABLE 4.9
PARTISANSHIP AND JUDICIAL SUPPORT FOR DEFENDANTS AND PROSECUTORS

When Defendant Supports	Votes with Defendant	Votes against Defendant
Democratic judges (86 votes)	25% (22)	75% (64)
Republican judges (55 votes)	14% (8)	86% (47)
When the State Supports	Votes against the State	Votes with State
Democratic judges (48 votes)	46% (22)	54% (26)
Republican judges	50% (7)	50% (7)

An examination of all judicial votes yields a similar, if weaker pattern (see table 4.8). With regard to holdings, 40 percent of Democrats and 45 percent of Republicans voted against polygraph evidence per se, disallowing even stipulated evidence. Democrats (43 percent) and Republicans (47 percent) similarly agreed to allow results if there was a stipulation. Where the partisan difference arises is in support for judicial discretion in polygraph admissibility (acceptance). Here, Democrats (16 percent) were twice as likely as Republicans (7 percent) to support accepting polygraph evidence as a matter of judicial discretion. This observation is congruent with dissent patterns: Democratic judges are more likely to vote for the liberalized admission of polygraph evidence.

While appearing counterintuitive to assumptions about liberals and law-and-order politics, polygraph evidence is more likely to be introduced by defendants rather than law enforcement. The parties of the judges were compared with the party seeking to admit polygraph results (see table 4.9). Judicial partisanship is

TABLE 4.10

REGION AND POLYGRAPH GATEKEEPING OUTCOMES

Region	Do Not Admit	Admit
Northeast	26 (93%)	2 (7%)
Midwest	38 (81%)	9 (19%)
West	37 (74%)	13 (26%)
South	36 (90%)	4 (10%)

Phi .196 (p = .096)

correlated with the party seeking admission of polygraph evidence. While both Democrats and Republicans are likely to reject polygraph evidence, Democrats are more likely to support defendant evidence (25 percent of Democrat votes, 14 percent of Republican votes). Democrats are also more likely to support state evidence (54 percent of Democrats, 50 percent of Republicans).

These three tests allow for triangulation signaling a general pattern for partisanship or attitudinal voting. While both groups of judges are likely to reject, Democrats are less likely to do so, apparently in support of defendant protection. Theoretically, this implies that attitudinal shortcuts can and will be utilized by judges. When the issue is multidimensional, however, the force of these shortcuts can be neutralized. It is possible that some judges simply supported the idea of stipulation agreements more generally.

Region

Another test of attitudinal factors is to look at regional patterns in gatekeeping outcomes. Variance with judicial outcomes for polygraph might be correlated with regional differences in attitudes as reflected by courts. There was no a priori specification of the direction of these attitudes. The regions correspond with known differences in political culture and other attitudinal measures, however, and they may mean that courts in certain parts of the country consistently approach polygraph evidence differently than courts in other regions.

Table 4.10 presents a regional breakdown of judicial outcomes. Cases admitting polygraphs were concentrated in the West (46 percent) and Midwest (32 percent). There was no discernible regional pattern in polygraph jurisprudence, however, even when regions were tested for state or defendant support.

While these regional patterns might suggest a trend toward a regional relationship with the admissibility of evidence or the support of a particular party, there was no statistically significant relationship between region and polygraph gatekeeping decisions. This might be due to the long timeline. Certainly the West of 1950 is different than the West of 1990 in terms of political attitudes. It might also mean there is little attitudinal relationship to state gatekeeping policy.

Explaining Polygraph Gatekeeping: A Role for Politics

Judicial outcomes for scientific evidence in general, and polygraph evidence particularly, are also partly correlated with the political positions of the actors and interests involved. Several political and organizational variables were examined to look at the role of the political system in gatekeeping outcomes. The organization and incentives of policy advocates and the lack of third-party validation clearly affected the admissibility of polygraph evidence in state supreme courts. In chapter 2, these factors were conceptualized as institutional and organizational political leverage. From a theoretical standpoint, these informational cues explain much about the way courts have responded to polygraph evidence. Political and organizational hypotheses are explored separately below.

Proponents

The party supporting admissibility is an important part of the institutional and organizational support for the science. Unlike DNA, where admissibility at trial was exclusively advocated by the prosecution, polygraph evidence was much more likely to be introduced by the defendant.[8] Defendants supported the admission of polygraph evidence in 106 cases. The state supported admissibility in 57 cases. Two cases had no proponent. This means that defendants argued for the validity and reliability of polygraph evidence in 65 percent of the cases coming before state supreme courts.

In table 4.11, the proponents of polygraph evidence are cross-tabulated with the judicial outcome. As with DNA, the state was generally more successful than the defendant. The state was able to gain admittance of polygraph evidence in one-third of the cases where it supported admissibility. Defendants were able to gain admittance of polygraph evidence in less than one-tenth of their cases. These findings can also be aggregated under the assumption that the parties were adversarial. By combining state wins with defendant losses, there were 114 cases where the state's argument prevailed and 49 cases where defendants' arguments prevailed. This means that overall the state was successful in polygraph admissibility

TABLE 4.11

SUCCESS OF PROPONENTS OF POLYGRAPH EVIDENCE (JUDICIAL OUTCOMES)

	Total Cases	*Admitted*	*Not Admitted*
Defendant supported	65% (106)	9% (10)	91% (96)
State supported	35% (57)	32% (18)	78% (39)

Chi-square 12.777 (p=.000)
Likelihood Ratio 12.196 (p=.000)
Phi .280 (p=.000)

cases 70 percent of the time, while defendants were successful 30 percent of the time. This pattern of party success is statistically significant. This implies that state variation in polygraph admissibility might depend on whether the state or the defendant was arguing for admissibility.

In table 4.12, the proponents of polygraph evidence are cross-tabulated with the judicial holding. The wins and losses are distributed differently across judicial holdings. For both defendants and states, their greatest loss came in the form of a general holding of inadmissibility. Defendants had all of the inadmissible per se rulings, however, and the state had the lion's share of holdings where the evidence was inadmissible even with a stipulation. States won more stipulated cases than defendants, implying that court gatekeeping policy variance might be due to the party arguing for admissibility. As argued in the DNA chapter, law enforcement and the state enjoy a significant institutional advantage in judicial proceedings.

States have more judicial success than defendants in gatekeeping decisions, even when the science is very controversial. A closer look at the state's success reveals that states overwhelmingly used the "middle-ground" admissibility of stipulated results to pursue admittance (see table 4.13). In all, 42 percent of the cases where the state was a proponent involved stipulated results. Compare this to only 8 percent of cases where the defendant was a proponent. This shows that

TABLE 4.12

SUCCESS OF PROPONENTS OF POLYGRAPH EVIDENCE (JUDICIAL HOLDINGS)

Party Supporting	Inadmissible Per Se	Inadmissible w/Stip.	Inadmissible in General	Inadmissible Need Stip.	Admissible Have Stip.	Admissible Jud. Discretion
Defendant	6	5	64	21	6	4
State	0	11	19	9	13	5

Likelihood Ratio 27.462 (p=.000)

Phi .402 (p=.000)

TABLE 4.13

PROPONENTS OF STIPULATED POLYGRAPH EVIDENCE

	Total Cases	No Stipulation	Stipulation
Defendant supported	65% (106)	92% (97)	8% (9)
State supported	35% (57)	58% (33)	42% (24)

Chi-square 25.940 (p=.000)

Likelihood Ratio 23.900 (p=.000)

Phi .399 (p=.000)

states have enjoyed success partly because of the introduction of stipulated admissibility. There are several reasons for this relationship. Stipulations overwhelmingly originate with the state in the processing of suspects. It is the state that will ask to administer a polygraph, and it is the state that will include a stipulation as part of the paperwork completed before the polygraph is administered. Thus, this ability of the state to obtain stipulations more readily is part of the story of the increase in stipulated admissibility. The state was taking the initiative and courts were open to the idea.

A final analysis of party support demonstrates a pattern of proponents over time. While there is no linear correlation of time with the party supporting polygraph evidence, state support increased during the period when stipulated results became part of the jurisprudence. It is clear that, on average, there was an increase in state support for polygraph admissibility from 1965 to 1985. There is also evidence of decreased state support in later decades.

Expert Testimony

While expert testimony was an important part of this analysis for DNA evidence, it had little role in the judicial processing of polygraph evidence. Very few polygraph cases (17 percent) discussed expert evidence. Courts seemed much more likely to assess relevant literature than to discuss experts at trial, if they opened the door to scientific analysis at all. Part of this is also due to the fact that many records provide no clues as to expert testimony. The battle of the experts in DNA evidence was clearly linked to the state's desire to secure admissibility. With regard to polygraph evidence, no aggressive campaign for admittance was undertaken on behalf of the most organized actor (the state), thus reducing the need for or use of expert testimony.

Policy Advocates (Amici Curiae)

Only two decisions recorded amicus briefs. These two decisions had a total of seven amicus briefs. In *People v. Barbara*, 255 N.W.2d 171 (Mich. 1977), the defendant was arguing for admissibility of polygraph evidence. Only one group of policy advocates supported this position: the American Polygraphers Association. Five other groups, including criminal defense groups, argued against admissibility. These groups included the American Civil Liberties Union, the state Prosecuting Attorney's Appellate Service, the Prosecuting Attorneys Association, the state Appellate Defense Office, and the state Association of Criminal Defense Lawyers. The court declined to allow polygraph evidence in Michigan.

In *People v. Angelo*, 666 N.E.2d 1333 (N.Y. 1996), the defendant was also arguing for admissibility. The sole amicus brief supported this position. The New York Association of Criminal Defense Lawyers argued that the defendant was entitled to present polygraph evidence as a matter of constitutional right. The court was not convinced, and it rejected polygraph admissibility.

While too small to draw conclusions about the role of amici curiae in scientific gatekeeping decisions, the findings are still informative. First, as with DNA, there appear to be the same three types of policy advocates. One type of policy advocate is internal to the legal system: public defenders and public prosecutors. The fact that three briefs represented public defenders and two briefs represented prosecuting attorneys is evidence of strong policy recommendations by legal personnel. Briefs representing law enforcement and criminal defense provide a picture of the values at stake in the use of this particular science. It is especially noteworthy that even among colleagues there was no consensus, with the Michigan Association of Criminal Defense Lawyers arguing against admissibility while the New York Criminal Defense Lawyers argued for admissibility. A second type of policy advocate is the industry supported by the science. One brief was filed by the American Polygraph Association. As expected, industry is supportive of the use of polygraph evidence in criminal prosecution. A third type of policy advocate is external to the legal system: public-interest groups. The American Civil Liberties Union filed an amicus brief against polygraph evidence.

Does policy advocacy matter? There is little to be drawn from these modest findings. In a broader policy context, it is clear that these provide cues to judges about the values and resources at stake in scientific gatekeeping decisions. Legal personnel, industry, and public-interest groups, as well as legal experts, clearly care about judicial gatekeeping outcomes. It is not surprising to see that the majority of the briefs represent advocates who have the most at stake in gatekeeping decisions: the attorneys who represent the accused and the attorneys who are charged with prosecution. As for explaining the overall patterns of acceptance of polygraph evidence, policy advocates clearly contribute to the picture of a system in conflict. As with DNA evidence, these findings highlight the larger conflict between legal personnel, with prosecutors and defense attorneys on different sides. It is not clear, however, that these are the prevailing views of all members of the legal community. Even the defense groups in Michigan and New York disagree with each other. Also, like DNA situation, industry is a staunch supporter of the inclusion of its products in the judicial system. It seems safe to say the contentions among policy advocates, even insiders such as legal personnel, will be reflected in judicial patterns.

Polygraphs, Congress, and the Department of Defense

Another source of institutional and organizational support is the availability of third-party reports on the acceptability of the evidence for legal purposes. Four key reports were repeatedly cited in polygraph cases. These reports represent both the political organization theory and the cognitive theory. A report with "legitimacy" will aid decision makers, and it takes political organization to commission and fund such reports. Government and law enforcement officials pushed for studies to offer "impartial" and expert analysis of polygraph evidence. The courts and those speaking to them were quick to use these reports to their advantage.

These reports arose from 1964 to 1984, with the following statements about polygraph evidence. In 1964, the Report of President's Commission on the Assassination of President Kennedy, cited in *Ballard v. Superior Court*, 410 P.2d 838 (Cal. 1966), included an extensive report on the use and reliability of polygraph exams. The commission took no position on polygraphs, and their report provided support for and against admissibility. Specifically the report noted the inability to test the accuracy of the test while at the same time characterizing the tests as "informative" (18). In 1976 the House Committee on Government Operations issued a report, which said, in part: "Although there is indication that efforts are being made to upgrade the training and educational requirements of polygraph operators, the committee finds that unproven technical validity of the polygraph devices themselves makes such efforts a meaningless exercise" (Committee on Government Operations 1976). In 1983 the Office of Technology Assessment issued a report claiming, "The instrument cannot itself detect deception.... The cumulative research evidence suggests that the polygraph test detects deception better than chance, but with significant error rates" (U.S. Congress, Office of Technology Assessment 1983). Finally, in 1984, the Department of Defense issued a more favorable assessment of polygraph testing, especially arguing for its general reliability: "There was no data suggesting that the various polygraph techniques have high error rates" (U.S. Department of Defense 1984).

These reports conceivably supply judges with cognitive cues, and it is reasonable to suspect that they have a significant bearing on judicial outcomes with regard to polygraph evidence. Nor are these reports equal in their persuasion. The more "neutral" the source of the report, the more legitimate and valuable the information is to the decision maker. The reports from congressional sources are to be more trusted than reports from the Department of Defense because Congress has no direct stake in the outcome. The Department of Defense makes extensive use of polygraphs, providing it with a significant interest in the scientific acceptance of polygraph evidence. The congressional reports were much more critical of polygraph evidence, and they were favorably received by judicial decision makers. Prior to 1976, courts were almost exclusively dependent on studies from polygraph insiders. An independent, third-party report was a welcome addition to the polygraph jurisprudence—even though these reports were not conclusive, they provided many courts with enough information to say polygraph evidence lacked general acceptance.

By revisiting table 4.1, it is possible to see a pattern of policy divergence and destabilization corresponding with the dates associated with these third-party reports. These shifts can be understood as the interaction of "middle-ground" development and inconclusive third-party support.

Multivariate Analysis of Polygraph Gatekeeping Decisions

After looking at the descriptive and bivariate statistics, it is obvious that certain variables may have an independent effect on the judicial outcome for polygraph evidence. This section of the analysis uses logistic regression to build a model of those

TABLE 4.14

LOGIT ANALYSIS OF JUDICIAL OUTCOMES FOR POLYGRAPH EVIDENCE

	B	S.E.	Wald	df	Sig.	Exp(B)
Type of analysis			18.790	2	.000	
Type of analysis (1)	−3.259	.753	18.717	1	.000	.038
Type of analysis (2)	−2.416	.747	10.473	1	.001	.089
Supporter (1)	−1.888	.572	10.899	1	.001	.151
Legal standard	.904	.423	4.572	1	.033	2.469
Constant	.692	.829	.697	1	.404	1.998

Chi-square (significance)		49.697 (.000)
−2 Log Likelihood	99.838	
Nagelkerke R-square	.49	
N= 163		

Prediction Table		Predicted		Percentage
		ADMIT	DO NOT ADMIT CORRECT	
Observed	ADMITTED	11	17	39.3
	NOT ADMITTED	2	133	98.5
				88.3 % Overall Correct

variables related to judicial gatekeeping and polygraphs. The dependent variable in this analysis is the judicial outcome for each case, coded as a binary dummy variable: zero for "not admitted" and one for "admitted." Three key variables are considered for multivariate analysis. First, the type of court analysis was a categorical variable coded 1, 2, or 3 depending on if the court used precedent/general rule, relevant scientific literature, or some other standard. Second, the party seeking the admissibility of polygraph evidence (supporter) was a categorical variable coded 1 or 2 for defendant supported or state supported. Third, the legal standard variable was coded with label 1, 2, 3, or 4 and placed in order of liberality (*Frye, Daubert*, FRE 702, and Other). It was modeled as an ordinal variable, with admissibility hypothesized to be positively related to the liberality of the legal standard.

A model of judicial outcomes for polygraph is provided in table 4.14. Multivariate analysis was performed using those variables described above after attempting analysis with a variety of significant and nonsignificant variables (such as region). The model in table 4.14 used 163 cases. Overall, the chi-square of 49.697 (.000) allows for the rejection of the null hypothesis (that all coefficients in the model are zero). It discards the null hypothesis by modestly predicting 39 percent of the "acceptance" cases and 98 percent of the "rejection" cases. Three variables have independent contributions to the outcome: the type of judicial analysis, the party seeking admissibility, and the legal standard used to decide the case. According to the estimated R-square, this model accounts for 44 percent of the total variance.

Multivariate Discussion

In terms of specific variables, the model supports several hypotheses. First, the type of court analysis clearly matters in this analysis of polygraph admissibility, independent of the legal standard or the supporting party. Each of the categorical variables was significant in its relationship to the outcome. Most notably, reasoning by the court using precedent or a general rule had a negative relationship to the outcome and decreased the odds of acceptance by 3.259. This is not surprising when the record reveals that precedent and the general rule stated that polygraph evidence was inadmissible. By contrast, reasoning using relevant scientific literature, while also having a negative relationship, decreased the odds of acceptance by only 2.416.

Second, the party seeking admissibility also had an independent relationship to the outcome, regardless of standard or type of analysis. Here, when defendants supported the admissibility of polygraphs, the odds of acceptance was decreased by 1.89. This implies, as the bivariate analysis demonstrated, that state-supported evidence is more likely to result in a finding of admissibility, independent of the legal standard or the supporting party.

Third, the legal standard is significantly related to the outcome. As expected from the bivariate analysis, legal standard has a positive relationship to the judicial outcome. As we move from a high standard of admissibility (such as the *Frye* general acceptance test) to a lower standard of admissibility (such as the FRE 702 relevancy test), admissibility becomes more likely.

More Clues to Gatekeeping Behavior

Polygraph jurisprudence is certainly more complex than DNA jurisprudence. Courts have struggled to develop policy in an uncertain scientific, legal, and political environment. Polygraph evidence possessed the trappings of a scientific profession, with journals, professionalization, and industrial applications. Traditional methods of establishing validity and reliability, however, such as testing and known rates of error, were not readily available for the judicial analysis of polygraph evidence. Polygraph evidence offered additional support for factual arguments in the courtroom, making its potential as a legal tool very promising. The lack of sufficient reliability and validity complicated a judicial determination about the usefulness or propriety of allowing polygraph evidence. Politically, polygraph evidence lacked a secure set of advocates, beyond the polygraphers themselves. While industry, government, and law enforcement made ample use of polygraphs, the judicial acceptance war was led by defendants. Conflicting expert assessments in third-party reports added to the scientific, legal, and political uncertainty.

While a difficult subject matter for state supreme courts, this analysis of polygraph evidence has confirmed important theoretical hypotheses, especially when discussed in conjunction with an analysis of DNA evidence. The theoretical foundation of this study was built on the concept of judges as human decision makers who frame and evaluate information. It was also argued that this cognitive process moves in the direction of efficiency: minimizing cognitive work by maximizing psychological satisfaction. Three empirical realities were expected to influence this cognitive process. On one level, the role of legal standards and the array of legal options are part of the framing and evaluation process. On a second level, the role of the decision makers' own political attitudes contribute to the framing and the evaluation process. On a third level, external factors, such as the political positions of proponents and policy advocates, as well as the availability of third party assessments, affect the framing and evaluation of judicial alternatives.

Theoretically, the overall pattern of polygraph jurisprudence is best understood as an interaction between the legal context and the political/institutional context. The stability of early polygraph jurisprudence is the result of a strict legal standard in operation. The destabilization of polygraph evidence is the result of a weakening in the legal context, due to the introduction of stipulated admissibility and the increase in political pressure. This is why polygraph jurisprudence can be called a victim of law and politics. For instance, the liberalism of the 1970s clearly pushed for more defendant-friendly proceedings. The professionalization of the polygraph industry and the widespread proliferation of polygraph application created pressure on the judicial system. Thus, with the advent of stipulated admissibility and the increase in systemic pressure for acceptance, the judicial system destabilized. The relative consensus after 1984 (thirty-four jurisdictions do not allow polygraph evidence in 2004) was helped along by critical reports. At this time, jurisdictions sorted themselves into stable policy outcomes either disallowing all polygraph evidence or allowing only stipulated results.

The analysis of judicial processing of DNA evidence and polygraph evidence has provided a foundation for the exploration of other science in the judicial system. The next chapter will examine the way jurisdictions have processed psychological syndrome evidence (rape trauma syndrome and battered wife syndrome). Legal standards and alternatives, attitudes, and political/institutional factors are expected to have a similar role in syndrome jurisprudence. Theoretically, it is expected that the framing and evaluation process will move toward psychological efficiency: minimum cognitive work for maximum cognitive satisfaction. Empirically, this process will be aided by these legal, attitudinal, and political/institutional factors.

Political Implications

For DNA evidence, it was concluded that legal standards do indeed control judicial processing of scientific evidence, with more liberal standards allowing for broader acceptance. For polygraph evidence, the additional role of legal alternatives in the judicial system was demonstrated. Politically, this means that the system will evolve with the development of innovative legal mutations. In terms of public policy, an adaptive legal system is clearly desirable. It is also comforting to see that the destabilization of the system did not seriously impair the overall goals of justice and sound policy development. It is also important to understand the ability of the system to absorb disagreements among the decision makers in the system. Finally, the ability of the legal system to evolve and experiment allows for optimal policy development. This was clearly evident with the experimentation of particular jurisdictions with stipulated admissibility. The overall ability of the judicial system to successfully sort a contested science is similarly reassuring.

As a system dominated by individual political actors, it is not disconcerting to find a partisan pattern in the judicial processing of scientific information. State supreme court judges are certainly legitimate political actors. It is important, however, to see that gatekeeping decisions are far from dominated by detectable political preferences. The pattern, while detectable, is not particularly strong for either DNA evidence or polygraph evidence. Judges are clearly responsive to the greater political context providing them with cues, information, and constraints—and these factors appear to override any political preferences. Thus, the system can be said to successfully sort scientific evidence in accordance with sound decision-making principles.

In terms of political and organizational resources, law enforcement clearly emerges as a powerful player in judicial processing of scientific information. DNA and polygraph evidence was significantly more likely to gain acceptance when supported by law enforcement. Furthermore, the fact that polygraph does its law enforcement work before it gets to the courtroom explains much about the difference in law enforcement pressure for DNA and polygraphy. While DNA acceptance was aggressively pursued, polygraph acceptance was less of a priority. Indeed, the development of stipulated admissibility of polygraph evidence can be seen as the extent of law enforcement pressure. The state pushed for admittance only with regard to a system where it held the advantage—obtaining stipulation agreements at the administration of the exam. It vehemently fought admission of non-stipulated results, especially results the defendant obtained from an independent polygrapher. It is no surprise to see organized insiders in the judicial process obtaining the most from the judicial system. Political scientists have long recognized the power differentials in organized, narrowly focused groups over diffuse, broad interests.

Third-party reports also provide important leverage in the judicial processing of scientific information. For DNA, the National Research Council provided sufficient support. For polygraphy, the Committee on Government Operations in the U.S. House of Representatives and the Office of Technology Assessment for the U.S. Congress provided sufficient reservations to discourage acceptance in the criminal courtroom. Two observations are important here. First, these studies are themselves the result of a political process. Political forces were at work to generate these reports. Second, the reports themselves were completed by independent actors with no direct stake in the outcome. The fact that these reports work to move the judicial system in a direction independent of direct political incentives implies that the system successfully sorted scientific evidence. The policy goal is to have outcomes of criminal proceedings result from empirically sound evidence. These two analyses of DNA evidence and polygraph evidence provide good reasons to feel that the system is working. Despite these analyses, legal commentators may feel that the system isn't working.

The exploration of psychological syndromes in the next chapter will provide additional tests for the role of politics in judicial processing of scientific information. As a "way of knowing," syndrome science is even less "scientific" than DNA evidence or polygraph evidence. It is also relatively new. It will be interesting to see if syndrome evidence is more controversial than DNA evidence or polygraph evidence. Syndrome evidence is expected to create different political cleavages than DNA or polygraph evidence. The policy advocates of syndrome evidence are expected to be more ideological than industrial, for syndrome evidence differentially affects ideologically defined values in the political system, particularly feminist values. One might expect divergence and attitudinal behavior to increase in this context. It will remain to be seen how legal and institutional constraints act on the processing of this particular type of science.

CHAPTER 5

SYNDROME EVIDENCE

SCIENCE ISN'T EVERYTHING

Science in the courtroom goes beyond forensic evidence and novel physiological measurements of emotional states. As another *way of knowing*, psychological syndromes present judges with even more complex decisional subject matter. To the uninitiated, psychological syndrome evidence may appear worlds apart from the lab science of DNA evidence or the wires and machines of polygraph evidence. Psychological syndromes, however, are considered scientific evidence in the judicial system and they are analyzed through the use of scientific admissibility jurisprudence (e.g., Giannelli and Imwinkelried 1993). The following analysis examines two gatekeeping puzzles for two syndromes: rape trauma syndrome and battered wife syndrome, the two most common syndromes in criminal law. These syndromes are used in criminal proceedings to provide information to the decision maker (usually a jury). Judges, as gatekeepers, must decide admissibility. Judges, as policy makers, are called on to allocate the political resource represented by syndrome evidence. As with the admissibility of DNA evidence and polygraph evidence, syndrome evidence similarly provides for political winners and losers.

The Science of Syndrome Evidence

Rape Trauma Syndrome

In standard textbook definitions a syndrome is a set of symptoms that together characterize a disorder. In 1974 Ann Burgess and Lynda Holmstrom published a yearlong study of 147 rape victims seeking assistance at the Boston City Hospital (Burgess and Holmstrom 1974). They established a pattern of symptoms common to rape victims and named the disorder "rape trauma syndrome." Rape trauma syndrome (RTS) is a subclass of posttraumatic stress disorders (PTSD), a set of disorders long recognized in war participants and victims of disaster. Symptoms are manifested as part of a long-term effect of a stressful event. With RTS, the stressor is rape.

Burgess and Holmstrom described rape trauma syndrome as a two-phase response. The initial phase, the acute phase, lasts from one to six weeks. According to Burgess and Holmstrom, victims in the acute phase exhibit either an expressive style or a controlled style. The expressive victim outwardly expresses fear, anxiety, and anger. In the controlled victim, these emotions are masked. The second phase, the reorganization phase, is a long-term adjustment phase. Victims may move or change their phone number. Victims also exhibit new phobias related to the rape, such as a fear of being alone.

In criminal trials involving rape or sexual abuse, rape trauma syndrome evidence is presented by an expert witness, usually a social worker or psychologist who counsels rape victims or studies rape behavior generally. Expert testimony will provide a description of typical victim behavior and/or a discussion of the symptoms as they appear in the complainant. Not all experts have personal knowledge of the victim in the case. Some may testify simply as general expert witnesses.

When it comes to thinking of rape trauma syndrome as *a way of knowing*, courts usually ask if RTS is a reliable and valid measure of the occurrence of a rape. At issue in many rape cases is whether the victim consented to the sexual encounter. The presence of syndrome symptoms can often be offered as "proof" that a rape occurred because a victim who consented to the encounter would not exhibit the symptoms, having not experienced the traumatic stress of forcible sexual assault. Courts have been skeptical of this approach. While a psychologist can reliably identify symptoms, several jurisdictions have pointed out that the information given to the psychologist is framed and provided by the victim. Furthermore, there is a problem of multiple causality—symptoms of posttraumatic stress disorder may arise from an event other than rape, much the way polygraphs may measure general anxiety rather than anxiety generated by a fear of detection. Even more problematic for many jurisdictions is the fact that rape trauma syndrome was developed as a therapeutic tool rather than a diagnostic tool. In other words, the syndrome is used to treat the problem, not to diagnose the problem. With this in mind, some courts note that while experts may consider the syndrome reliable for therapeutic reference it does not follow necessarily that it is reliable for legal purposes, where the goal is to discern questions of fact and questions of law.

Despite these shortcomings, rape trauma syndrome has found several uses in the legal system. Evidence of rape trauma syndrome is almost always presented to bolster the prosecution of a rape crime. RTS has been used to explain counterintuitive behavior of a complainant, particularly delay in reporting and subsequent contact with the accused. RTS has also been used to demonstrate a lack of consent; in other words, the presence of the symptoms signals the sexual encounter may not have been consensual. Finally, RTS has been used as evidence to prove that a rape occurred, even when physical evidence was not conclusive.

From a judicial perspective, rape trauma syndrome is considered by state supreme courts to be reliable for some purposes but not for others. First, most courts agree that RTS testimony is reliable to explain the counterintuitive behavior of a rape victim, such as delay in reporting and inconsistent recall. Second, many courts allow RTS testimony to rebut a rape case where the defense is consent. Courts, however, generally do not support the use of RTS testimony to prove that a rape occurred. Thus, part of the variance in state supreme court gatekeeping policy is related to the multidimensional use of rape trauma syndrome testimony.

Battered Woman Syndrome

As mentioned above, a syndrome is a set of symptoms that together characterize a disorder. In 1979 Lenore Walker published a study of 110 victims of spousal abuse. She established a pattern of symptoms common to victims of spousal abuse and named the disorder "battered woman syndrome" (BWS). In a follow-up study of a more representative sample of 435 women, she published a more complete view of the syndrome (Walker 1984). According to Walker, a battered woman is one who has been subjected to repeated psychological or physical abuse by a man with whom she has had an intimate relationship.

Walker described battered woman syndrome as a three-phase pattern of abuse that results in a distinct emotional change in the victim. The initial phase, the tension-building phase, is characterized by small episodes of abuse. The second phase, the acute battering phase, is characterized by violence that is out of control. The third phase, the contrition phase, is characterized by loving behavior undertaken by the batterer to repair the damage in the relationship. This cycle is then repeated. Walker observed that battered women are often trapped in these relationships for economic or psychological reasons, and many of them exhibit the psychological trait of "learned helplessness"—a condition where they feel that they can no longer control the abuser's behavior.

Like rape trauma syndrome, battered woman syndrome evidence is presented by an expert witness, usually a psychologist or a social worker who directs a women's shelter. Battered woman syndrome, however, is almost always presented by a defendant who is accused of murder. The line of testimony is developed to allow the battered woman to claim that she murdered in "self-defense." Expert testimony provides a description of the cycle of abuse to explain the perspective and subsequent behavior of a battered woman so as to conform to the legal requirements for self-defense (namely that reasonable person would agree that harm was imminent).

When it comes to thinking of battered woman syndrome as *a way of knowing*, courts usually ask if BWS is a reliable and valid explanation of behavior and perception of danger. The primary use of BWS is exculpatory: a battered woman who is accused of murder will use BWS to bolster a claim of self-defense, particularly to provide a holistic picture of the perceived imminent danger. BWS testimony is also used by prosecutions to rebut attacks on a complainant's credibility or to explain behavior, especially recantations. Critics of the use of battered woman syndrome testimony have charged two shortcomings. First, BWS, when used in a self-defense case, cannot explain why a woman attacks her abuser when she does (Faigman 1986). Second, when BWS is used on behalf of the state, it can be very prejudicial to the defendant. Both criticisms challenge the ability of BWS as a reliable way to know if a situation involves battering.

The Politics of Syndrome Evidence

As with DNA and polygraph evidence, three political processes have a role in judicial outcomes: the process creating resources for a legal contest (institutional and organizational politics), the process creating the calculus for legal decision making (legal standards), and the cognitive framing process used by the decision makers themselves (attitudes). Unlike forensic DNA or polygraph evidence, however, syndrome evidence is primarily used in and developed for nonlegal applications. Both rape trauma syndrome and battered woman syndrome are therapeutic tools for psychologists in a patient relationship. A cottage industry of syndrome expertise has arisen for legal purposes, however, and the same set of experts often testify in high-profile cases.

These two syndromes present an interesting opportunity for institutional and organizational comparison because one, rape trauma syndrome, is overwhelmingly used by the prosecution while the other, battered woman syndrome, is overwhelmingly used by defendants. This variance in institutional support might presuppose that RTS would enjoy greater judicial support. These syndromes, however, are not politically defined in terms of law-and-order concerns. RTS and BWS are more alike than different when it comes to broader political forces. That is, both of them are on the same side of the feminism debate. Thus, while a law-and-order analysis would grant RTS greater political strength, a gender-based analysis would link the fate of these tools to the greater political battle over women's rights. Gender concerns lend organizational and institutional support to both syndromes—funding studies, training experts, even passing majority rules legislation to support them. The interesting question is whether law-and-order politics can overpower gender politics or vice versa. Because RTS evidence unites prosecutors (institutional and organizationally more powerful than defendants) and feminist politics, we might expect RTS to enjoy greater support than BWS.

Obviously, these syndromes present yet another test for the attitudinal model. What is not clear is the extent to which law-and-order values will trump social values as ideologically defined. In other words, when faced with a choice between a decision that supports simultaneously the accused and conservative values, which choice will win out in a conservatively oriented judge? Similarly, when a decision is both pro-feminism and anti-accused, how will a liberal-leaning judge decide?

Finally, the legal standard for admissibility becomes murky with syndrome evidence. Some courts find it reliable, even as they say it is too prejudicial to a defendant. Because syndrome evidence is not "lab" science, it is not obvious to all jurisdictions that it should be analyzed under scientific admissibility standards. Some jurisdictions did not use conventional legal tools in their admissibility standards. The choice of a legal calculus is clearly a political choice, and it can often lead to a different outcomes.

As with polygraph evidence, we might expect to see a variety of nuanced holdings rather than a straight yes-or-no decision. With polygraphy, I argued that the wiggle room created by the scientific ambiguity allowed for wiggle room in judicial outcomes. Syndrome evidence is similarly situated in terms of scientific vulnerability. This vulnerability can lead to a plurality of judicial decisions even within the same jurisdiction. In other words, rather than a simple yes or no, it is expected that syndrome evidence will lead to a variety of holdings and possible legal mutations.

The Gatekeepers and Syndrome Evidence

When faced with the question of admitting syndrome evidence as scientific evidence, courts had multiple possible responses. These responses can be placed roughly along a spectrum. For both syndromes, the spectrum goes from per se rejection and rejection for some purposes, to acceptance for other purposes and general acceptance. More explicitly, battered woman syndrome was either rejected, allowed for explaining behavior only, or allowed more generally at the discretion of the trial judge. Similarly, rape trauma syndrome was either rejected per se, rejected to prove that a rape occurred, rejected to rebut a defense of consent, admissible to rebut a defense of consent, admissible to explain behavior, or allowed more generally at the discretion of the trial judge. Inherent in these decisions is a judicial determination of scientific validity and reliability.

General Findings

The analysis reported relatively few cases where state supreme courts ruled on the admissibility of battered woman syndrome and rape trauma syndrome as a question of scientific reliability and validity. In general, there were forty battered woman syndrome evidence cases and thirty-one rape trauma syndrome evidence cases. These cases were collected using a state-by-state LexisNexis search of all high court cases containing the phrase "rape trauma syndrome" or "battered woman syndrome."[1] Each case was then briefly read to isolate those cases in which the court ruled on the validity/reliability and admissibility of syndrome evidence. These cases are not a sample; they compose the complete picture of court activity. Each case was coded for the holding with regard to syndrome evidence. The following descriptive statistics break this down by year, state, and holding.

Judicial Outcomes

In tables 5.1 and 5.2, the initial pattern of acceptance and rejection is shown across time for rape trauma syndrome and battered woman syndrome. Several observations are immediately forthcoming. First, the overwhelming majority of decisions admitted battered woman syndrome (68 percent) and rape trauma

TABLE 5.1

ADMISSIBILITY OF RAPE TRAUMA SYNDROME EVIDENCE
IN STATE SUPREME COURTS

RTS Admissible	Year	RTS Inadmissible
KS	1982	MN, MN
	1983	
CA, KS, MT	1984	KS, MO
AZ	1985	
KS, MD	1986	MT
CO, IN	1987	WA, WY
	1988	OR, PA, WV
IA, SD	1989	
MA	1990	NY
	1991	
NY, SD, IL	1992	NC
SC	1993	NM
	1994	
CT, KS	1995	MD
	1996	
	1997	
	1998	
	1999	
VT	2000	
	2001	
	2002	KS
Total cases 17 (55%)		14 (45%)

syndrome (55 percent). Second, these decisions are spread fairly evenly across the same twenty-year span. These cases begin to surface in appellate jurisdictions in the early 1980s, only a short time after their inception. Unlike DNA or polygraph evidence, there is no converging or diverging pattern across time. Rather, the controversy is much more consistently distributed across time.

When the mean judicial outcomes for rape trauma syndrome and battered woman syndrome are charted over time, there is a relatively steady nature of the controversy with regard to rape trauma syndrome and battered woman

TABLE 5.2

ADMISSIBILITY OF BATTERED WOMAN SYNDROME EVIDENCE
IN STATE SUPREME COURTS

BWS Admissible		BWS Inadmissible
GA	1981	OH, WY
	1982	
ND	1983	
NJ, WA, WA	1984	
	1985	TX
KS, SC	1986	
KS	1987	
WA	1988	KS
	1989	MN
OH, KY	1990	
PA	1991	
OK, RI, SC, SD	1992	MS
CT	1993	
	1994	WA
MI, VT	1995	MT
	1996	WV
MN	1997	
GA, WY	1998	
KY	1999	WY, WY
MA, MA	2000	NV
	2001	
	2002	
	2003	
CA, CA	2004	
Total cases 27 (68%)		13 (32%)

syndrome. The mean holding changes slightly in the 1990s when syndrome evidence was re-debated after much criticism in the legal and psychological literature. Contrary to expectations, rape trauma syndrome has more variance and is therefore empirically more controversial than battered woman syndrome.

TABLE 5.3

HOLDINGS FOR SYNDROME EVIDENCE

Holding for Rape Trauma Syndrome Evidence	State Supreme Courts (n)
RTS is inadmissible per se	16% (5)
RTS is inadmissible to prove that rape occurred	23% (7)
RTS is inadmissible when consent is the defense	7% (2)
RTS is admissible when consent is the defense	23% (7)
RTS is admissible to explain behavior	16% (5)
RTS is admissible (judicial discretion)	16% (5)

Holding for Battered Woman Syndrome	State Supreme Courts (n)
BWS inadmissible per se	18% (7)
BWS admissible to explain behavior only	25% (10)
BWS admissible (judicial discretion)	58% (23)

In table 5.3, judicial outcomes are further disaggregated to show more detailed patterns in the debate among jurisdictions. Like polygraph evidence, syndrome evidence yielded multiple responses in the courts, presenting a spectrum of outcomes from "inadmissible *per se*" to "admissible at judicial discretion." As the tables demonstrate, most jurisdictions choose a middle ground position for RTS. For RTS, six positions were possible: inadmissible per se, inadmissible to prove a rape occurred, inadmissible to rebut a defense of consent, admissible to rebut a defense of consent, admissible to explain behavior, or allowed more generally at the discretion of the trial judge. On the rejection side, the most popular holding for inadmissibility was to prove that a rape occurred. The largest group (23 percent) did not think RTS was reliable or valid to prove that a rape actually happened. The judicial reasoning was framed as a multiple causality problem: the posttraumatic stress disorder symptoms could have been caused by an event other than the alleged rape at issue.

On the acceptance side, the most popular holding for rape trauma syndrome admissibility was admissibility to rebut a claim of consent. The largest group (23 percent) supported the use of RTS to rebut a claim of consensual intercourse. These courts reasoned that RTS symptoms were reliable enough to indicate that the victim was traumatized by the encounter, thereby implying that it was not consensual. Furthermore, in consensual cases RTS is often used to explain subsequent behavior of a victim. In contrast, a few jurisdictions (7 percent) rejected this argument, holding that RTS was not admissible when consent was the defense. These jurisdictions reasoned that the testimony of an expert and the use of the term "rape trauma syndrome" implied a scientific reliability beyond the court's assessment of the true capabilities of RTS. Other courts (16 percent) were also likely to admit RTS evidence to explain counterintuitive behavior by a complainant or victim. Rape victims typically will not behave as a jury might expect. For example, victims often delay reporting or continue their contact with a known perpetrator.

For battered woman syndrome, three judicial outcomes were recorded: inadmissible per se, admissible to explain behavior only, and admissible with judicial discretion. Most jurisdictions (58 percent) allowed the admission of BWS at the discretion of the trial judge. In other words, if the expert is qualified, the testimony is admissible. BWS in almost all of these cases is part of a murder trial where the defendant is a woman claiming self-defense. In this context, BWS is used to explain the perception of imminent danger. As one jurisdiction pointed out, "Part of the very root of the problem with courts accepting BWS is the so called 'reasonable man' standard. How a reasonable man may perceive the danger is quite possible wholly different from how a reasonable woman may perceive the danger" (Concurring Opinion, *Commonwealth v. Dillon*, 598 A.2d 963, 971 [Penn. 1991]). And another jurisdiction put it this way: "The very essence of the admissibility of testimony on battered woman syndrome is predicated on the assertion that a battered woman does not react to outside stimulus in the same way that a reasonable person would react; i.e. that the perceptions of a battered woman are different from those of a reasonable person" (*Fielder v. State*, 683 S.W.2d 565, 592 [Tex. 1985]). These courts clearly feel that BWS is a reliable explanation of the psyche of a "battered woman."

A minority of courts (18 percent) were skeptical of the reliability and validity of BWS, especially in establishing a psychological state of perceived imminent danger. Early decisions, such as in Ohio in 1981, found that BWS was too new to be generally accepted among the relevant scientific community. Other decisions would not admit BWS until it was established that the woman defendant was a battered woman (Texas, 1985). Likewise, Montana (1995) did not admit BWS because no foundation was laid that the case involved a battered woman. Kansas, on the other hand, was admitting BWS. In a 1988 decision, however, Montana said that BWS could not be used as a defense—for example, BWS evidence cannot be used to excuse a defendant. Still, other jurisdictions did not allow for novel applications of BWS. For instance, in 1994, Washington did not allow BWS evidence to be admitted in an unusual case involving the murder of a child.

Some jurisdictions (25 percent) allowed BWS testimony to explain general behaviors of battered women. Again, this was to disabuse lay jurors of misconceptions of battered women—for instance, why a woman would remain with someone who constantly mistreated her. This testimony is particularly useful when there is no police record of domestic violence in the relationship.

Black Robes in Action

As with other scientific evidence, courts vary in the amount of energy they have expended on syndrome jurisprudence. With regard to rape trauma syndrome, fifteen jurisdictions had only one RTS evidence case (Arizona, California, Colorado, Connecticut, Illinois, Indiana, Iowa, Massachusetts, Missouri, New Mexico, North Carolina, Pennsylvania, South Carolina, Vermont, and Washington). Five jurisdictions had two RTS cases (Maryland, Minnesota, Montana, New York, and South Dakota). Kansas had unusually prolific rape trauma syndrome gatekeeping activity, revisiting the issue of admissibility in six separate decisions. With regard to battered woman syndrome, the numbers are similar but the states change. Fourteen states had only one BWS decision (Connecticut, Florida, Michigan, Mississippi, Nevada, New Jersey, North Dakota, Oklahoma, Pennsylvania, Rhode Island, South Dakota, Texas, Vermont, and West Virginia). Seven jurisdictions had two BWS cases (Georgia, Kentucky, Massachusetts, Minnesota, Montana, Ohio, and South Carolina). Kansas is again distinguished with three BWS cases. Washington and Wyoming each revisited the question of BWS admissibility four times.

The most interesting set of jurisdictions are those that have decisions for and against the same syndrome evidence (see table 5.4). These states represent a dynamic movement in their jurisprudence. These are jurisdictions where initial and subsequent decisions about syndrome evidence varied. Unlike DNA evidence and polygraph evidence, there is no dominant pattern in individual state jurisprudence. Neither syndrome can be said to move toward or away from

TABLE 5.4
JURISDICTION PATTERNS FOR DYNAMIC JURISDICTIONS

Rape Trauma Syndrome	Admissible (Inadmissible)
Kansas	1982 (1984) 1984 1986 1995 (2002)
Maryland	1986 (1995)
Montana	1984 (1986)
New York	(1990) 1992

Battered Woman Syndrome	Admissible (Inadmissible)
Kansas	1986 1987 (1988)
Minnesota	(1989) 1997
Ohio	(1981) 1990
Washington	1984 1984 1988 (1994)

acceptance. Moreover, there is no chronological pattern where certain years or decades indicate a shift in individual state jurisprudence.

In reality, much of the intra-jurisdiction behavior is due to decisions that are answering different questions about the limits of syndrome evidence. For example, Kansas, traditionally open to syndrome evidence, and indeed offering the first appellate decision in support of rape trauma syndrome, foreclosed on a novel application of BWS to a non-intimate relationship. Thus, as the applications of syndrome evidence mutate, judges must readjust and reexamine the reliability and validity of the syndrome evidence.

Other intra-jurisdictional behavior does indicate a change in precedent due to increasing scientific acceptance. The prime example is the reexamination of battered woman syndrome in Ohio. In 1981, the Ohio court had found BWS to be too new to warrant admission under the *Frye* standard. When it reexamined the issue in 1990, however, the growth of relevant literature and the increased general acceptance among those in the psychological field had convinced the judges to allow BWS testimony. Similarly, Minnesota moved from rejection to acceptance of BWS from 1989 to 1997.

Other jurisdictions initially opened the gates to the evidence, only to cabin it in later decisions. For instance, Maryland admitted RTS under a liberal standard in 1986 and later restricted its use in 1995. Kansas also had to adjust its RTS rules in subsequent decisions.

Disagreements on the Bench

Historically, state supreme courts rarely feature judicial debate in the form of concurring and dissenting opinions (Jaros and Canon 1971). Because of this, the magnitude and direction of such debate is important in understanding the judicial processing of scientific information. Here, with syndrome evidence, judicial debate provides more information about the character of disagreement among colleagues in the same jurisdiction faced with the same facts.

Among judicial colleagues, syndrome evidence is much more controversial than DNA evidence or polygraphs. The decisions and opinions are annotated in table 5.5. Thirteen (42 percent) rape trauma decisions and twelve (30 percent) battered woman cases had concurring or dissenting opinions where justices disagreed with the majority ruling on syndrome evidence. These are high percentages compared to DNA and polygraph cases (where closer to 10 percent had concurring and dissenting opinions about the science). For RTS, justices on both sides of the admissibility issue wrote separate opinions. For BWS, separate opinions were only written in cases where the court admitted the evidence. As with other forms of scientific evidence, the debate centers on whether the science in question is valid and reliable for the purported purpose.

Looking closer at rape trauma syndrome, there is evidence that justices were often likely to disagree with their colleagues on the admissibility of RTS.

TABLE 5.5

SUMMARY OF JUDICIAL DEBATE FOR SYNDROME EVIDENCE

Rape Trauma Syndrome		Court Majority	Concurring or Dissenting Argument
1984	Kansas	Admissible	(Concur) Writes to rebut defense, supports RTS (Dissent) Wants to overturn *Marks*, too prejudicial
1984	Montana	Admissible	(Dissent) RTS far from scientific exactitude
1986	Maryland	Admissible	(Dissent) RTS cannot scientifically prove a rape occurred
1987	Colorado	Admissible	(Concur) Remember this is a narrow holding (Dissent) Should only admit RTS when it satisfies *Frye*
1989	South Dakota	Admissible	(Dissent) Invades province of the jury (Dissent) Unfairly gives victim the aura of truth
1990	Massachusetts	Admissible	(Concur) Different application, should not admit here
1992	North Carolina	Admissible	(Dissent) RTS is no more inflammatory than other expert evidence
1992	South Dakota	Admissible	(Dissent) Should not admit
1993	South Carolina	Admissible	(Dissent) Should only admit to explain behavior
1982	Minnesota	Inadmissible	(Dissent) Data and research support RTS admissibility
1986	Montana	Inadmissible	(Dissent) Too picky, should admit RTS and expert testimony
1987	Washington	Inadmissible	(Concur) Should not close the door on RTS
1988	Pennsylvania	Inadmissible	(Dissent) Should admit to aid the jury (Dissent) Should admit

(*continued*)

TABLE 5.5

SUMMARY OF JUDICIAL DEBATE FOR SYNDROME EVIDENCE (CONTINUED)

Battered Woman Syndrome		Court Majority	Concurring or Dissenting Argument
1984	Washington	Admissible	(Dissent) Effectively silences the prosecution
1988	Washington	Admissible	(Dissent) Too prejudicial, but okay for self-defense case
1989	Pennsylvania	Admissible	(Concur) Should not have addressed BWS
1990	Kentucky	Admissible	(Dissent) Expert overstepped and not qualified
1990	Ohio	Admissible	(Concur) Remember limits of BWS
1991	Pennsylvania	Admissible	(Concur) BWS must be admissible (Concur) BWS important to dispel behavioral myths
1992	Oklahoma	Admissible	(Dissent) Not clear that BWS has much to do with defendent's case
1995	Michigan	Admissible	(Dissent) Not the same as child sexual abuse syndrome
1995	Vermont	Admissible	(Concur) BWS expert is a "disguise"
1997	Minnesota	Admissible	(Concur) Possible prejudicial effect, be careful
2000	Massachusetts	Admissible	(Dissent) BWS is not part of this case
2000	Nevada	Admissible	(Dissent) BWS not relevant
2004	California	Admissible	(Dissent) Misapplied statute, strongly disagrees

Over half (nine) of the RTS rulings admitting RTS were decided by split votes. A quick perusal of the arguments demonstrates that not all justices were convinced of the scientific reliability and validity for the purported legal purposes. Almost one-third (four) of the RTS decisions rejecting RTS evidence also experienced bench disagreements. The arguments of these dissenting justices demonstrate that many justices were convinced of the reliability and validity of RTS in direct contradiction of the majority of their colleagues. These debates are particularly interesting when we consider that these colleagues heard the same evidence applied to the same fact situation in the same jurisdiction.

Looking at battered woman syndrome, justices were also likely to disagree at high rates. About one-third of all BWS cases (thirteen) experienced significant judicial debate. These concurrences and dissents are much less explicit about the scientific validity of BWS, often talking more generally about the limits of self-defense doctrines. They clearly felt, however, that battered woman syndrome is problematic for criminal justice purposes.

The Consequences of Judicial Analysis

Unlike polygraph and DNA evidence, courts tended not to review the literature or the expert testimony. Rather, most cases were isolated balancing analyses in which a certain level of validity was assumed but weighed against other judicial concerns for a fair trial. Only a few cases undertook a thorough analysis of the state of the art of syndrome evidence. This places syndrome evidence at a cognitive disadvantage because judges did not or could not make use of the profession's information on the usefulness of syndrome evidence.

Overall, the pattern of judicial outcomes for syndrome evidence appears to follow a pattern similar to polygraph evidence. First, there is controversial debate about the reliability and validity of the use of syndrome evidence for certain legal applications. Second, the debate "mutates" as legal personnel adapt new strategies and find subsequent novel applications for the syndrome evidence. Thus, when it comes to the characterization of the "debate" it must be realized that part of the debate is about the degree of reliability and validity for specific applications. This is new to our discussion because polygraph evidence did not involve mutated applications—the core of polygraph uses remained the same: the detection of deception. Syndrome evidence, however, has been used to inform (e.g., typical behaviors of syndrome victims) as well as to serve as a detection tool (e.g., to detect the occurrence of an event, such as battering or forcible rape). Clearly the detection purposes of syndrome evidence are most controversial. Using syndrome evidence to determine an issue of fact—such as the occurrence of a rape or the presence of a battering relationship—is what appears to put syndrome evidence on shaky ground. As judges acknowledge, the science of syndrome evidence was not developed to detect as much as to treat syndrome symptoms.

As judges decide which ways of knowing are reliable and valid for legal purposes, three sets of factors of acting are relevant to their decisional patterns. The theoretical approach of this analysis sees the court as a set of cognitive decision makers who are using cues from politically created variables. The variance in these politically constructed cues is a key part of the story of state-to-state gatekeeping variation. One set of factors is the role of legal standards and the legal context on judicial determinations of reliability and validity. Liberal standards should correlate with liberal admissibility standards. Similarly, mutations in the system can change the weights used in legal calculus. Another set of factors is the role of the decision makers' political attitudes or belief systems. Judicial decisions necessarily allocate power and as such are political decisions. As political actors in a political system, judges may have preferences for certain political winners and losers, and their preferences may influence their ability to frame and evaluate alternative courses of action. A third set of factors is the role of political organization and political institutions in providing resources for judicial outcomes. Each set of factors will be examined below.

Explaining Syndrome Gatekeeping: Legal Considerations

Does the law determine judicial outcomes? In the analysis of DNA evidence and polygraph evidence, there was moderate evidence of a legal pattern in judicial outcomes. Generally, the question centers on the role of calculi in the determination of the outcome. In the context of scientific reliability and validity, three judicial standards are commonplace. By way of review, there are essentially three standards used in the judicial admittance of scientific evidence: the relevancy standard (state versions of the Federal Rules of Evidence Section 702, hereinafter FRE 702), the general acceptance standard (*Frye* standard), and the *Daubert* standard (four potential criteria for a judicial decision based on the *Daubert* case). In terms of syndrome evidence, the legal standard at work in a jurisdiction is expected to influence judicial outcomes. The more liberal the standard, the more likely it is the jurisdiction will find the science acceptable. The more difficult the standard, the more likely scientific evidence will be rejected.

The problem with syndrome evidence is that not all courts treated it as scientific evidence.[2] While some courts analyzed RTS and BWS under the traditional scientific acceptance standards, other courts used a balancing test or a method not associated with conventional determinations of reliability or validity—even though these decisions implicitly include a determination of reliability. In table 5.6, the legal standards are presented for syndrome evidence. A little less than half of RTS and BWS cases used the three conventional standards for scientific analysis; in the majority of cases, courts did not use the usual scientific calculi normally employed by the judicial system. This provides further evidence of the degree to which the law affects the judicial outcome. It is also interesting to note that there is more room to change the legal standard used by a court in this non-lab kind of science. Some courts simply do not define it as a science: they frame and analyze the issue as something other than scientific reliability and validity.

In table 5.6, the judicial outcomes for rape trauma syndrome and battered woman syndrome are also presented. First, while there is a pattern among the conventional legal standards, it is not statically significant, perhaps because of the small

TABLE 5.6

LEGAL STANDARDS AND JUDICIAL OUTCOMES FOR SYNDROME EVIDENCE

Legal Standard	Cases	RTS Admitted	RTS Not Admitted
Frye or "general acceptance"	6 (20%)	2	4
Daubert (one of four criteria)	0 (0%)	0	0
FRE 702	6 (20%)	4	2
Other	19 (60%)	12	7

Phi = .333 (p = .248)

Legal Standard	Cases	BWS Admitted	BWS Not Admitted
Frye or "general acceptance"	5 (12%)	3	2
Daubert (one of four criteria)	1 (2%)	0	1
FRE 702	7 (17%)	6	1
Other	27 (68%)	18	9

Phi = .286 (p = .352)

sample size. Of those using *Frye* analysis, two decisions admitted and four rejected. Of those using FRE 702 tests, four admitted and two rejected. While statistically inconclusive, the pattern is consistent with the findings in DNA and polygraph evidence: the more conservative the standard, the less likely the science will be admitted. Second, the syndrome evidence analyzed under other standards is more likely to be admitted, especially when judged in comparison to *Frye* analysis. For both RTS and BWS, two-thirds of the decisions using a nonconventional standard were likely to admit the evidence. Indeed, some dissenters on admitting courts called for a *Frye* analysis and maintained the position that syndrome evidence would not pass a *Frye* standard. These numbers demonstrate a correlation between judicial standard and admissibility in the judicial processing of syndrome evidence.

As with polygraph evidence, there is a need to examine the greater legal context to fully understand the role of law in judicial outcomes. With polygraph evidence, the role of the law was changed because of the advent of a middle ground of "stipulated" admissibility. This greatly lengthened the life of polygraph evidence in the criminal justice court. With syndrome evidence, the role of the law has to do with the character of the case rather than procedural innovations. Some cases are more likely to open the judicial gates for syndrome evidence than others, based on the nature of the legal claim. This portion of the legal model suggests that the legal

TABLE 5.7
THE ROLE OF LEGAL FACTS IN SYNDROME ADMISSIBILITY

RTS Factual Situation	Cases	RTS Admitted	RTS Not Admitted
Consent is the defense	13	62% (8)	38% (5)
Not a consent case	18	55% (10)	45% (8)

Phi .060 (p=.739)

BWS Factual Situation	Cases	BWS Admitted	BWS Not Admitted
Self-defense is the defense	30	63% (19)	37% (11)
Not a self-defense case	10	80% (8)	20% (2)

Phi −.154 (p = .330)

facts of the case can similarly determine the outcome. Indeed it is true for syndrome evidence: the legal facts of the case greatly mattered in judicial outcomes.

There was a significant factual pivot point for both RTS evidence and BWS evidence. In table 5.7, these factual parameters are compared. For RTS, the gate was opened further when the defendant in a rape case claimed that the sexual contact was consensual. These "consent cases"—cases where consent was the real issue—were more likely to allow the admission of RTS evidence: 62 percent of consent cases admitted while only 55 percent of non-consent cases admitted. For BWS, the court was less likely to admit when the defendant, a woman who was on trial for murdering a man with whom she had had an intimate relationship, claimed the murder was committed in self-defense. BWS in self-defense cases was rejected (37 percent) at higher rates than BWS in non-self-defense cases (20 percent).

These pivot points have characteristics similar to the use of stipulated evidence in the polygraph cases. As non-admitting jurisdictions point out, these factual situations do nothing to change the validity or reliability of syndrome evidence. Hence, some jurisdictions concede that syndrome evidence can explain behavior, but they feel that it cannot be used to decide whether a course of action was legally justified. In other words, both RTS and BWS pivot points depend on the kind of defense offered. In neither set of facts is there an issue about who did what, only whether it was justified. Syndrome evidence is called on to tease out the truth in a situation where the difference is a state of mind—clearly a relevant use of syndrome evidence because it involves psychological information. It is also this use of syndrome evidence that requires a finding of validity and reliability. Can the presence of RTS mean that this sexual encounter was not consensual? Can the presence of

BWS mean that this confrontation presented an imminent danger in the mind of the accused, thereby justifying murder in self-defense?

By allowing certain fact situations to dictate the gatekeeping function, courts have provided a way for syndrome evidence to find its way into the courtroom. And, like the polygraph cases allowing stipulated polygraphs, it can be argued that some admissibility of syndrome evidence is more psychologically satisfying than foreclosing all admissibility or a finding of complete reliability and validity. We get the feeling that judges are treating syndrome evidence like polygraph evidence: while these syndromes are not "conclusive," neither are they "useless." In terms of policy implications, the use of these middle-ground approaches can be beneficial for the system as a whole. If the goal of the criminal justice system is to process defendants, the use of syndrome evidence certainly aids in that process.

As with polygraph evidence, not all courts used a middle-ground approach. They preferred the logical consistency of a scientific determination of validity and reliability: if RTS cannot be used in fact situation A, it cannot be used in fact situation B. These were the courts on the extremes of judicial holdings: courts ruling syndrome evidence inadmissible per se or courts ruling syndrome evidence admissible at the discretion of the trial court judge.

Explaining Syndrome Gatekeeping: Democrats, Republicans, and the West

In this analysis, testing for attitudinal relationships is a test of partisan and regional relationships to judicial outcomes. With regard to DNA evidence and polygraph evidence, partisan patterns were present in judicial voting patterns. A partisan pattern was expected with regard to syndrome evidence. Here, the same two-step process was used to obtain judicial partisanship data. First, the names of the judges serving on the cases were obtained. For most courts, these names were found on the opinions. Second, most judges were labeled with a partisan label (Democrat or Republican). These labels were obtained either by a direct label (in the *Judicial Yellowbook*) or an indirect label extrapolated from the party of an appointing governor.

There were 208 named judges on the state supreme court cases spanning twenty years of syndrome evidence jurisprudence. This yielded a total of 294 judicial votes on syndrome evidence. Of these votes, 176 were identifiably partisan; 94 were cast by Democratic judges and 82 were cast by Republican judges. These identifiably partisan judges provided some support for attitudinal differences.

First, the partisan pattern does not distinguish between RTS and BWS (see table 5.8). Rather, even though Republicans and Democrats acted differently, each group showed a consistent level of support for each syndrome. In all, 62 percent of Republicans and 50 percent of Democrats supported the admissibility of rape trauma syndrome evidence. Similarly, 78 percent of Republicans and

TABLE 5.8

PARTISANSHIP AND JUDICIAL VOTES TO ADMIT SYNDROME EVIDENCE

Rape Trauma	RTS Inadmissible	RTS Admissible
Democrats	50% (22)	50% (22)
Republicans	38% (10)	62% (16)

Chi-square .877 (p = .349)
Likelihood Ratio .473 (p = .491)
Phi −.112 (p = .349)
One-way ANOVA F .863 (p = .356)

Battered Woman	BWS Inadmissible	BWS Admissible
Democrats	50% (25)	50% (25)
Republicans	22% (10)	78% (36)

Chi-square 8.260 (p = .004)
Likelihood Ratio 7.085 (p = .008)
Phi −.293 (p = .004)
One-way ANOVA F 8.849 (p = .004)

50 percent of Democrats supported the admissibility of battered woman syndrome evidence. For BWS, the results were statistically significant. These patterns seem to indicate that Republican justices are significantly more likely than Democratic justices to support admissibility of syndrome evidence. Furthermore, while Democrats were split, Republicans clearly favored acceptance.

Second, the votes were further disaggregated to compare partisanship with winners and losers. The parties of judges were compared with the party (prosecution or defense) seeking to admit polygraph results. For rape trauma syndrome, these numbers are the same as above, for every case analyzed was a case where the state was seeking to introduce evidence. For battered woman syndrome (table 5.9), Republicans highly favored the evidence whether admissibility was argued by the defense (77 percent voted with the defendant) or the prosecution (80 percent voted with the state). Democrats likewise demonstrated little change regardless of whether admissibility was argued by the defense (49 percent with defendants) or by the prosecution (50 percent voted with the state). When the cases are combined, this means that Republicans voted in support of the defense in 59 percent of the cases. This is slightly more than the Democrats who sided with the defense in 49 percent of the cases.

TABLE 5.9

PARTISANSHIP AND JUDICIAL SUPPORT FOR DEFENDANTS
AND PROSECUTORS IN BWS CASES

When Defendant Supports BWS	Votes with Defendant	Votes against Defendant
Democrat judges (35 votes)	49% (17)	51% (18)
Republican judges (31 votes)	77% (24)	23% (7)

When the State Argues for BWS	Votes against the State	Votes with State
Democrat judges (16 votes)	50% (8)	50% (8)
Republican judges (15 votes)	20% (3)	80% (12)

TABLE 5.10

PARTISANSHIP AND DISSENTING OPINIONS

Dissenting Opinions	Rape Trauma Syndrome		Battered Wife Syndrome	
	DEMOCRATS	REPUBLICANS	DEMOCRATS	REPUBLICANS
We should not admit	2	3	4	3
We should admit/ Trial court discretion	1	0	0	0

Third, an examination of dissent patterns provides only a slight partisan pattern (see table 5.10). Almost all dissenters, regardless of party, argued against the admissibility of syndrome evidence. Of those dissenting justices whose partisanship could be ascertained, five (two Democrats and three Republicans) wrote against the admissibility of RTS and seven (four Democrats and three Republicans) wrote against the admissibility of BWS. One Democrat wrote a dissent in favor of syndrome evidence (RTS). The only notable part of this pattern is the fact that there is a trend toward Democratic protest of BWS admissibility, a syndrome supported at higher percentages (78 percent) by Republicans. Thus, this dissent pattern, while inconclusive, is at least consistent with above findings.

This analysis demonstrates that a moderate partisan difference in support for syndrome evidence, regardless of whether the state or the defendant was arguing for admissibility. While it makes sense under law-and-order values for Republican justices to support rape trauma syndrome as a tool of the prosecution, it is not clear which schema is at work in a conservative support of battered woman

TABLE 5.11
REGION AND GATEKEEPING OUTCOMES FOR SYNDROME EVIDENCE

Region	Do Not Admit RTS	Admit RTS
Northeast	3 (38%)	5 (62%)
Midwest	5 (36%)	9 (64%)
West	4 (57%)	3 (43%)
South	1 (50%)	1 (50%)

Phi .180 (p = .800)

Region	Do Not Admit BWS	Admit BWS
Northeast	1 (14%)	7 (86%)
Midwest	4 (36%)	7 (64%)
West	7 (58%)	5 (42%)
South	1 (14%)	7 (86%)

Phi .410 (p = .088)

syndrome. It might be that Republican judges are simply better educated. It is clear, however, that even though these syndromes have significant political overtones in terms of feminism, there are apparently other political attitudes at work among Republican justices. The nearly even split among Democrats on rape trauma syndrome is perhaps a bit more intuitive when we think of the fact that RTS jurisprudence requires choosing between women's rights and the rights of the accused—both traditionally liberal values.

These three tests allow for triangulation signaling a general pattern for partisanship or attitudinal voting. While both groups of judges are likely to accept, Democrats are *less* likely to do so. Theoretically, this implies that attitudinal shortcuts can and will be utilized by judges. When the issue is multidimensional (such as choosing between law-and-order politics and feminist politics), however, the force of these shortcuts can be neutralized. Thus, while attitudes may play a role in judicial outcomes, they are not the whole story, at least not in terms of conventional conceptualizations of ideological values. Even so, attitudes and partisanship may partially account for the variance in state supreme court syndrome gatekeeping jurisprudence.

Region

One other set of data was investigated to look for constraints on judicial attitudes. Regional differences in syndrome admissibility did arise (table 5.11).

As with polygraphs, Western states distinguished themselves from other regions. For both syndromes, there was a trend toward Western rejection of admissibility. For rape trauma syndrome, the West was the only region where a majority (57 percent) of cases did not admit. For battered woman syndrome, the West was once again the only region where a majority (58 percent) of cases did not admit. While not statistically significant, this regional variance may indicate a trend toward regional differences reflecting attitudinal differences. For rape trauma syndrome, the Western propensity to reject RTS evidence means that defendants faired better in the West because the state supported admissibility in all RTS cases.

For battered woman syndrome, region was further analyzed in terms of the party seeking admissibility to see if this pattern held. Here, defendants and the state alike were rejected at higher numbers in the West. This implies that the activity of the West may not be related to attitudes toward defendants or the state. Rather, there may be a different value at stake, such as a skepticism of the science or an antifeminism sentiment.

Explaining Syndrome Gatekeeping: A Role for Politics

The explanation of judicial gatekeeping outcomes for syndrome evidence extends beyond legal and attitudinal factors. Syndrome evidence also implicates a set of political and organizational factors similar to DNA and polygraph evidence. The relative political positions of proponents and opponents of syndrome evidence, as well as the availability of expert witnesses and the presence of third party validation, are political factors associated with judicial outcomes for syndrome evidence. Theoretically, it is important to remember that these factors provide "cues" to cognitive decision makers on the state supreme courts implying that they can increase the political leverage of a particular type of evidence. Courts, as policy makers, rely on the information supplied by these factors to make a decision regarding the acceptability of syndrome evidence for judicial purposes. It is also important to remember that political forces create these factors.

Proponents and Appellant Status

By way of review, legal scholars have long recognized that appellants are favored in judicial proceedings (e.g., George and Epstein 1992). For syndrome evidence, almost all appeals were from defendants. Defendants were the appellants for thirty (96 percent) of the thirty-one rape trauma syndrome cases. Similarly, defendants were the appellants for thirty-six (90 percent) of the forty battered woman syndrome cases. For rape trauma syndrome, the state was appealing a trial court rejection of RTS evidence, and it won that case. For battered woman syndrome, the state was appealing defendant use of BWS, and the prosecution won two and lost two. Overall, defendants had significant success as appellants in RTS cases (42 percent) and in BWS cases (68 percent). This is especially noteworthy because appeals are routine in criminal cases.

In terms of the party seeking the admissibility of syndrome evidence, there was a distinctive pattern (see table 5.12). For RTS, the state (100 percent) was the only party seeking admissibility at the state supreme court level. For BWS, most of the cases were defendant supported (80 percent). For both syndromes the state was more successful than the defense in getting syndrome evidence admitted. The state was able to win the majority of the time it was seeking admissibility for both RTS (58 percent success) and BWS (75 percent success). This implies that state support for the evidence was a strong indicator of success. Defendants, however, were likewise successful the majority of the time they sought admissibility (66 percent success with BWS). Thus, it cannot be said that the party seeking admittance is necessarily correlated with judicial outcomes.

When the proponent figures for all BWS cases were combined, the defense (57 percent) won more cases than the state (43 percent). This is counterintuitive since we would expect prosecutors to enjoy an organizational advantage in the system. The BWS decisions were further broken down by specific holding to understand more of the story (see table 5.13). Here, it is noticeable that the state

TABLE 5.12

SUCCESS OF PROPONENTS OF SYNDROME EVIDENCE

Rape Trauma	Total Cases	RTS Admitted	RTS Not Admitted
Defendant supported	0	0	0
State supported	100% (31)	18 (58%)	13 (42%)

Battered Woman	Total Cases	BWS Admitted	BWS Not Admitted
Defendant supported	80% (32)	66% (21)	34% (11)
State supported	20% (8)	75% (6)	25% (2)

TABLE 5.13

SUCCESS OF PROPONENTS OF BATTERED WOMAN SYNDROME
(JUDICIAL HOLDINGS)

Party Supporting	Inadmissible Per Se*	Admissible Behavior Only	Admissible
Defendant	7	8	17
State	0	2	6

Chi-square 2.283 (.319)
Likelihood Ratio 3.622 (.164)

never lost with a per se inadmissibility holding. Rather, the two state losses were due to an overreach of expert testimony. In other words, the state was told that it was permissible for a BWS expert to testify generally about BWS behavior, but it was not allowed the latitude of presenting any opinion about the particular woman involved. Thus, it appears that the defendant was still at a disadvantage when attempting to present BWS evidence.

Overall, syndrome evidence is somewhat more likely to be admitted when it is supported by law enforcement. These findings are consistent with the findings for DNA and polygraphy. By triangulation, it is now possible to postulate that the party seeking admissibility has a notable relationship to the success of admissibility. State support is apparently a strong political and organizational advantage. State prosecutors typically have more resources and staff than defense attorneys, especially court-appointed defense attorneys. They may also have more skills and experience with this particular kind of evidence. In terms of the puzzle of state variance in gatekeeping decisions, it is likely some of this variance is partially an effect of the political leverage of the party seeking admissibility.

Expert Testimony

Experts for syndrome evidence, unlike DNA or polygraphy, do not have law enforcement as their primary client base. Rather, these are individuals who work with a certain group of individuals, namely, rape victims and victims of domestic violence. Unlike forensic lab techs and polygraphers, the experts in these fields are not primarily concerned with law enforcement applications.

Many of the syndrome cases named the syndrome witness and described her qualifications. These experts were social workers, counselors, or directors of women's shelters, as well as medical personnel who work with victims. Often, the cases included a dispute about the relative expertise of these individuals and the proper qualifications for an expert opinion.

Also unique to syndrome evidence is the expert's relationship with the woman in question or similarly situated women. These experts necessarily bring empathy (bias?) to their testimony, even when they are only testifying generally about the recognized manifestations of the syndrome. Counselors, social workers, and medical doctors are clearly sympathetic to victims of rape or spousal abuse. This is quite different from DNA experts and polygraphers who, while likely to foster favorable attitudes toward law enforcement, are very likely to hold unfavorable attitudes toward the person they are testifying about (except for the "friendly" polygraphers sometimes employed by defendants). As "victims," those who are the subject of RTS or BWS testimony generally have the support and empathy of the expert testifying about their case.

These "human" elements of expert testimony are important when we consider judges as cognitive decision makers. These policy makers must make decisions based on the "cues" they receive. Because of inherent empathy, the system necessarily fosters a favorable presentation of the information—favorable to the person who is the subject of such testimony—in direct contrast to DNA and polygraph evidence.

All the RTS cases discussed expert testimony, with 58 percent admitting the evidence. Only a portion (65 percent) of the BWS cases discussed expert testimony. Those BWS cases that discussed expert testimony had higher admissibility rates (69 percent) than those cases that did not discuss expert testimony (36 percent).

Why would courts be more likely to discuss expert evidence for RTS cases than for BWS cases? In DNA evidence, expert evidence was front and center. In polygraph cases, it was hardly noted (17 percent of the cases mentioned the expert). To repeat an earlier observation, the battle of the experts with DNA was very important to the state—prosecutors aggressively pursued admissibility. The same logic applies to RTS and BWS—the state has a very high stake in the admissibility of RTS and certain kinds of BWS evidence. With RTS, which was exclusively supported by the state, it was especially important to secure judicial acceptance and to have the court address the parameters of expert testimony. Thus, RTS, like DNA, represents an aggressive campaign for admittance on behalf of the most organized political actor (the state). Compounding the problem is the inability of the defense to find an expert to argue *against* RTS evidence. A similar pattern was detected with regard to expert testimony and DNA evidence.

BWS, on the other hand, was largely seen as the enemy of the prosecution—and expert testimony was often stifled in pretrial motions. Under this reasoning, because BWS was not the exclusive tool of law enforcement, and indeed because it was much more likely to be used by a defendant, there was little incentive for the state to push the acceptance of expert testimony. Therefore, there was no expert testimony to be dissected and discussed in many BWS cases. The state could argue against its admissibility early on in the trial process, often convincing judges not to allow any testimony. With no testimony, there is no expert to discuss. This is a continued advantage to law enforcement officials who may be able to challenge BWS admissibility because the exact limits of expert testimony have not been thoroughly established by jurisprudence. Indeed the majority of BWS cases where the evidence was held to be "generally admissible" were cases in which the court was saying the defendant *should have* been allowed to introduce BWS testimony. Thus, the exact parameters of expert testimony and the conditions under which the testimony would be considered reliable and valid were yet to be litigated.

One last point about battered woman syndrome and expert testimony: the state tended to fight BWS at the legal level rather than the trial level. Rather than a battle of the experts, prosecutors sought to have BWS testimony limited because of matters of law (mostly self-defense jurisprudence), not because of matters of scientific accuracy. Thus, rather than attack the science of BWS testimony, they attacked the legal permissibility of such evidence, often successfully keeping BWS testimony from surfacing at the trial stage.

Policy Advocates

As with DNA evidence and polygraph evidence, syndrome evidence has its own set of supporters and detractors. A few of these policy advocates surfaced in "friend of the court" briefs. Six decisions recorded amicus briefs for syndrome evidence (see table 5.14). These six decisions had a total of eight amicus briefs. With regard to RTS, there was only one set of supporters: California Women Lawyers. This group argued for the admissibility of RTS evidence, apparently as part of a general advocacy of women's rights. There were two briefs filed against RTS evidence. In a California case, the State Public Defender's Office filed amici curiae to argue against the admissibility of RTS. In a Pennsylvania case, the Defender's Association of Philadelphia also argued against admissibility. With regard to BWS, there were several supporters, all of them part of the legal community. The National Clearinghouse for the Defense of Battered Women, the Northwest Women's Law Center, and the Washington Association of Criminal Defense Lawyers argued in support of BWS. The Kansas County Prosecutors and District Attorneys Association filed the sole state supreme court amicus brief against BWS.

TABLE 5.14

AMICUS BRIEFS IN SYNDROME CASES

RTS Decision	Holding	Amici to Admit	Amici to Reject
California	Inadmissible	California Women Lawyers	State Public Defender's Office
Pennsylvania	Inadmissible		Defender's Assoc. of Philadelphia

BWS Decision	Holding	Amici to Admit	Amici to Reject
Kansas	Inadmissible		Kansas County Prosecutors And District Attorneys Assoc.
Pennsylvania	Admissible	National Clearinghouse for the Defense of Battered Women	
Washington	Admissible	Washington Association of Criminal Defense Lawyers NW Women's Law Center	
Washington	Admissible	NW Women's Law Center	

It is apparent that policy advocates for and against syndrome evidence arise solely from those internal to the criminal justice system. Every amici curiae is associated with legal personnel. There is also a sense that some of these policy advocates have broader ideological interests, namely the protection of women and women's rights. The California Women Lawyers, National Clearinghouse for the Defense of Battered Women, and the Northwest Women's Law Center are all attorney groups specifically targeted toward the protection of women. The presence of these groups provides evidence of the gender-based policy considerations inherent in RTS and BWS. The table also makes it apparent that the three cases with amicus briefs against syndrome evidence are the only cases in the chart where the evidence was held inadmissible. Conversely, the three cases with only supporting amicus briefs were cases where the evidence was held admissible. While the sample is too small to generalize, the pattern of congruence between advocacy and outcomes is noteworthy. Policy advocates obviously provide information and cues to decision makers. Justices are aware when certain players, especially repeat players such as the defender's association or a district attorney's association, lobby for a particular policy with regard to scientific evidence. With regard to syndrome evidence, all four of these repeat-player attorney groups received favorable decisions.

Third-Party Validation

One source of third-party validation was repeatedly cited in several syndrome evidence cases. Unlike government reports or the work of independent councils, syndrome evidence was squarely left with the professionals. Several decisions referred to the American Psychological Association's *Diagnostic and Statistical Manual of Mental Disorders* (3rd ed.) (DSMIII), published in 1987. This professional manual, which is periodically updated, describes all the mental disorders recognized by the psychiatric profession. Among commonly recognized disorders are posttraumatic stress disorders, which represent the class of disorders behind the conceptualization of rape trauma syndrome and battered woman syndrome. While expected in the BWS cases, no explicit references were made to the DSMIII. Thus, this analysis focuses on the use of the DSMIII in RTS cases.

As with the reports of the National Research Council in DNA evidence cases, state supreme courts were quick to discuss the *Diagnostic and Statistical Manual of Mental Disorders*. Ten of the thirty-one rape trauma syndrome decisions mentioned the DSMIII. Of these decisions, three mentioned the listing of rape as a possible stressor causing PTSD, while seven only discussed the general acceptance of posttraumatic stress disorders as a class. The courts acknowledged the professional acceptance of the syndromes. This did not, however, guarantee acceptance. Half of these decisions did not admit RTS evidence as valid and reliable. These decisions rejected syndrome evidence as clinically useful but legally inappropriate because they could not be used to identify the actual occurrence of a rape. The other half of the decisions admitted RTS; these opinions used

DSMIII recognition of posttraumatic stress disorders as proof that syndrome evidence was generally accepted as reliable in the psychiatric profession.

Unlike the third-party tipping point associated with the 1996 NRC DNA report, judges in syndrome cases were left to make the leap from clinical to legal use, using a manual written for professional rather than jurisprudential purposes. Unlike the reports of the National Research Council, DSMIII was not a report focused on legal applications. Rather, it was a professional manual of reference information developed by the profession that uses this science for therapeutic purposes. It might be expected that a report focused specifically on legal applications would be much more of a tipping point. Surely such a report would provide a formidable threshold for any jurisdiction seeking to ignore its conclusions. Even so, when faced with uncertainty, courts referenced the manual as a source of information. Thus, syndrome evidence is yet another example of the potential role of third-party reports in judicial gatekeeping decisions.

More Clues to Gatekeeping Behavior

This analysis of the judicial processing of syndrome evidence produced results similar to the previous analyses, allowing for a sense of triangulation.[3] As with polygraph and DNA evidence, syndrome evidence provided an opportunity for further testing of key legal, attitudinal, and political hypotheses. Syndrome evidence was chosen for three reasons: (1) the jurisprudence of rape trauma syndrome and battered woman syndrome had sufficient variation in judicial outcomes among state supreme courts; (2) syndrome evidence is analyzed by the judicial system as scientific evidence, and it is recognized as scientific evidence by the legal community (see Giannelli and Imwinkelried 1993); and (3) it is sufficiently unlike DNA and polygraphy to represent an entirely different "set" of scientific jurisprudence, thus providing a robust context for testing hypotheses about the judicial processing of scientific information.

The choice of syndrome evidence did not disappoint as a further proving ground for the major theoretical concerns of this work. First, with regard to the legal model, it was again demonstrated that the standard used to measure legally viable science has an important role in judicial outcomes. Specifically, as with DNA and polygraphy, syndrome evidence was less likely to be admitted in a jurisdiction employing a stricter standard. Second, with regard to the attitudinal model, syndrome evidence was also treated differently by the judges of each political party. Specifically, syndrome evidence was much more controversial among Democrats than Republicans, especially for BWS evidence. This partisan pattern held regardless of whether admissibility was supported by the state or the defendant. Third, with regard to political and organizational support and its role in syndrome evidence, syndrome evidence was more likely to succeed when it was supported by the state and certain repeat players within the legal system. Similarly, syndrome evidence admissibility also depended on the court's deference to third-party recognition of syndrome evidence. Thus, taken together, the findings support the role of all three models in judicial outcomes.

Political Implications

The key political question of this analysis is what do these theoretical relationships, and their proof in the findings, mean in terms of political winners and losers? While this particular syndrome evidence would appear to support the notion of a kinder, gentler legal system that has embraced gender concerns, the reality is that this evidence was not treated any differently than any other scientific jurisprudence. Legal, attitudinal, and political/organizational factors contributed to the success. Of particular interest is the fact that rape trauma syndrome, as the first syndrome to break the barrier, was introduced by the most powerful actor in these gatekeeping decisions—the state. Rape trauma syndrome arguably paved the way for syndrome evidence in the judicial system, and it was pushed by those actors with the most legal resources: the prosecution. Also of interest is that Democrats, conventionally perceived as champions of gender values, are not to be credited with the judicial acceptance of syndrome evidence. Rather, it was the Republicans who supported syndrome evidence in greater numbers than the Democrats, regardless of whether the state or the prosecution was seeking admissibility.

The state and the judges themselves (by defining legal standards and the weight given to particular legal facts) exercise the most control over scientific admissibility. Professionals and practitioners can only control judicial determinations of scientific reliability and validity if they are invited by the political system to study the issue (as with DNA and polygraph evidence). Syndrome evidence did not gain much power from the profession operating outside the legal and political system. Furthermore, other legal players were able to weigh in with some evidence of success as amici curiae.

Moreover, the relative power of these actors again implies that scientific merits alone do not determine judicial outcomes. The system requires time and action on the part of those who manufacture the cues used by judicial decision makers. For instance, there is a clear pattern of increasing acceptance across time—at least for *some* science. This means that there are several larger political forces at work in state supreme court gatekeeping decisions. In other words, the potential of the science to change the legal and political game puts forces into motion to shore up the reliability and validity of scientific innovation. It also puts forces into motion to question these new "ways of knowing." Thus, science requires political support (broadly defined as all relevant factors) to reach a threshold of acceptance. This can explain why syndrome evidence, while arguably less objective than DNA or polygraphy, was able to claim impressive support by the judicial system. Obviously the judicial system does not process knowledge in the same way that science does.

CHAPTER 6

GATEKEEPERS AND THE POLITICS OF KNOWLEDGE

The Politics of Gatekeeping Decisions

Politics is very much a part of judicial decisions to admit or reject novel scientific evidence. Indeed, political variables have a great deal more to do with gatekeeping than scientific variables. The preceding analysis examined state supreme court jurisprudence with regard to forensic DNA evidence, polygraph evidence, and psychological syndrome evidence. Each of these "ways of knowing" is a very peculiar stranger to admit to our judicial city. DNA evidence was the type of stranger initially received with mixed outcomes, only to find gatekeepers ultimately converging on acceptance. Syndrome evidence continues to experience conditional acceptance in a "well, it depends . . ." kind of jurisprudence. Polygraph evidence has completely fooled the gatekeepers, providing no sense of predictability and forcing other political actors (legislatures) to usurp gatekeeping power. But that is another story . . .

This work demonstrates the role of politically defined variables in judicial validation of particular types of science. This is possible because the gatekeepers are human, and as such, they are subject to the limitations of human cognitive capacity. Judges cannot *know* if a science is valid and reliable. They must rely on others to inform them. This information is filtered by legal standards and by political attitudes, each of which is also politically created. Therefore judicial gatekeeping decisions are political decisions. There is no elaborate algorithm to sort science into admissible and inadmissible categories. The task is simply part of a political process, and the outcome is a policy. This general observation is an important part of the story of state variation in judicial gatekeeping policy.

Where are the interesting stories in this policy-making context? They are found in the legal context, the attitudes of the policy makers themselves, and the larger institutional and political context. The puzzle of scientific gatekeeping decisions is unraveled with consideration of legal, attitudinal, and institutional/political variables.

Gatekeeping and the Legal Context

Legal standards and legal facts are important in judicial processing of scientific information. Variation in legal standards and legal facts account for much of the state variation. For three types of science (DNA evidence, polygraph evidence, and syndrome evidence), legal doctrines were a significant part of the story. As hypothesized, conservative admissibility standards, such as the *Frye* rule, were much more likely to result in judicial rejection. Indeed, for DNA, 80 percent of decisions rejecting DNA evidence as valid and reliable were in jurisdictions using the conservative "general acceptance" standard. For polygraphy, where there was variance in legal standards, liberal standards were still associated with admissibility and conservative standards were associated with inadmissibility. For syndrome evidence, where the evidence was sometimes analyzed under other, unconventional doctrines, the conservative standard resulted in rejection two-thirds of the time while unconventional standards resulted in acceptance two-thirds of the time.

Two of the case studies demonstrated an additional role for legal facts in the explanation of state variance. Polygraph and syndrome evidence were much more vulnerable to scientific attacks as a result of the role of subjective human conclusions in the production of the evidence. Polygraph science is largely dependent on human actors for reliability and validity at several stages in the testing. Syndrome evidence was developed as a therapy, where patient perception is more important than facts. These subjective elements (which, while present, are less evident in forensic evidence) challenged gatekeepers who were inclined to feel that the evidence could have some usefulness in legal deliberations. As such, the jurisprudence developed important factual situations that in and of themselves determined judicial outcomes. With regard to polygraph evidence, agreement to admit the results (stipulation) was likely to determine admissibility in several jurisdictions. With regard to syndrome evidence, the type of defense (*consent* for RTS and *self-defense* for BWS) determined the admissibility of the evidence. Unlike other legal arguments, these fact situations were inherently related to determinations of reliability and validity. The allowance of stipulated results assumes that both sides agreed to a passing level of reliability and validity in the results of a polygraph exam. Syndrome evidence was similarly assumed valid and reliable in certain fact situations—to a point. Thus, these particular legal facts are part of the story of scientific admissibility and variation in state policy making.

Gatekeeping and Judicial Preferences

Scientific evidence is not considered in a political vacuum. Rather, policy makers recognize the political implications of the admissibility of scientific evidence. Part of the mystery of the gatekeepers is accounted for when we consider that state courts can vary in their political goals due to individual judicial ideology or regional differences in policy preferences. Gatekeeping decisions create political winners and losers, and scholars of conventional models of judicial behavior will not be surprised to find that the framing and evaluation of a decision is, in part, correlated with the gatekeeper's measurable political attitudes. As hypothesized, Democratic judges behaved significantly differently from Republican judges in all three case studies. With regard to DNA evidence, Republican judges (90 percent) were much more likely than Democratic judges (77 percent) to vote in favor of admissibility (a decision that also meant a vote in favor of the state). With regard to polygraph evidence, Democrats (25 percent) were more likely than Republicans (14 percent) to favor defendant-introduced evidence. With regard to syndrome evidence, Republicans (62 percent) sided with the state more often than Democrats (50 percent) to allow RTS evidence. In an interesting finding, Republicans (78 percent) also showed stronger support for BWS evidence than Democrats (50 percent), regardless of the party arguing for admissibility. Thus, for syndrome evidence, while both types of judges favored admissibility, Democrats were *less* likely to do so. In general, much of the variation in state policy might be a function of court variation in judicial partisanship.

Another manifestation of attitudes or constraints on attitudes was detected in terms of regional differences in the judicial treatment of novel scientific evidence. For DNA evidence, courts in the Northeast were twice as likely as courts in other regions to reject. Even in multivariate analysis, the Northeast was a significant variable. For polygraph evidence, cases admitting the evidence were more concentrated in the West than in other regions, yet no statistically significant relationship was determined between region and judicial outcome. Likewise, for syndrome evidence, the West was more likely to reject than other regions, yet it lacked a statistically significant relationship. It was hypothesized that the small-n for polygraph admissions and syndrome evidence generally may have limited the ability of statistical tests to detect patterns.

Gatekeeping and Political Advantage

Scientific evidence does not stand alone before the courts of the land. The political strength of proponents and opponents is an important part of judicial outcomes. The variation in state gatekeeping behavior is partly due to the political leverage of the science in the case. In criminal proceedings, the prosecution has several advantages. The state is more successful than the defendant in obtaining favorable gatekeeping decisions—whether for admission or rejection. The analysis of DNA evidence demonstrated that the prosecution has a higher success rate (71 percent) on appeal than the defendant (17 percent). The DNA evidence also demonstrated that the state has more attorneys working on a case than the defense. Furthermore, with regard to DNA, the state was able to provide more and better expert witnesses than the defense. As the DNA analysis demonstrated, the prosecution has a political and organizational advantage.

With regard to polygraph evidence, the state continued to maximize its use of the relevant technology. Here the battle was engaged prior to prosecution, and polygraphs were used to induce confessions. Prosecutions also front-loaded the admissibility question by obtaining stipulations for the evidence. Furthermore, the state's arguments again prevailed in overall admissibility wins and losses, where the state won 70 percent of the cases. Syndrome evidence likewise confirmed the state's advantage with scientific admissibility. When the state sought admissibility, it won the majority of its cases. Overall, scientific evidence is much more likely to be admitted when it is supported by law enforcement. Similarly, law enforcement is able to keep unwanted scientific evidence out of the judicial system on a fairly consistent basis. The admissibility of scientific evidence is significantly correlated with its relationship to the prosecution.

What about other political actors? Where were they in the process? The analysis of policy advocates in the judicial system found a few surprises. First, formal advocacy (amici curiae) was largely limited to those within the legal system. Larger political interest groups did not get involved in scientific admissibility questions—even when it came to gender issues and syndrome evidence. Second, even among legal personnel, there were relatively few cases with amici briefs. Why would policy advocates maintain a low profile in jurisdictional gatekeeping decisions? Part of the answer may lie in separation of powers. There are other paths to policy changes in judicial decision making. The most obvious answer is that policy advocates could (and did) successfully lobby state legislatures to admit certain kinds of scientific evidence by statute. While it is not part of this analysis, it is worth noting that many states have codified admissibility rules for DNA, polygraph, and syndrome evidence. No doubt an analysis of legislative policy making in this area would uncover significant contributions by policy advocates.

One of the key findings in this analysis is the importance of third-party validation. The story of DNA jurisprudence is almost entirely explained by two National Research Council (NRC) reports. Courts were willing to incorporate these independent assessments into their decisions. The reports were endowed with authority and legitimacy, and the NRC's conclusions about DNA validity and reliability virtually disposed of all other challenges. Polygraph jurisprudence had the opposite problem. Here, conflicting reports were generated—causing further confusion in judicial outcomes. There was no expert consensus to provide the courts with guidance. Likewise, syndrome evidence did not provoke third-party validation tailored to the needs of the judicial system. Rather, courts had only an unannotated professional manual for reference. Overall, it is suspected that third-party reports can and do affect judicial outcomes, especially when they target the legal uses of the science.

Solving the Mystery

The original puzzle was created by the variation in state judicial treatment of scientific evidence. Presumably, the actual science did not vary from case to case. How do we account for the difference? First, it was necessary to describe judicial outcomes in terms of the cognitive processing of the decision makers. By defining the question as a decision undertaken under conditions of uncertainty (thereby recognizing the dependence of the judges on cues), we were able to theorize about those factors that likely affected judicial outcomes. These factors arise from conventional understandings of the relationships among law, attitudes, and institutions in judicial outcomes, and they proved very helpful in explaining our puzzle. Law and legal standards were clearly correlated with judicial variation. Attitudes and partisanship were related to judicial votes. Repeat players and appellant status, as well as third-party reports, were very much a part of the success (or failure) of scientific evidence in the state judicial systems.

While the conventional factors explained much of the puzzle of gatekeeping decisions, the analysis also demonstrated that questions of scientific admissibility are particularly vulnerable to the power of the state (institutional strength) and the power of third-party validation (judgments external to the judicial process). Gatekeeping decisions are distinguished from other types of judicial policy-making decisions because the increased uncertainty results in a judicial system much more dependent on external cues. The power of the state eclipses the power of defendants to capitalize on this process.

The conventional factors also allowed for a measurable degree of prediction about judicial outcomes and gatekeeping decisions. If we know the legal doctrines and relevant legal facts, if we know partisan composition, and if we know the state's position and the availability and content of third-party reports, we could begin to predict what state supreme courts are likely to do with this science. Think for a minute about several other novel scientific questions we could have included in this analysis. Perhaps we could go on to examine the jurisprudence with regard to hair analysis, or spectrograph (voiceprints) analysis, or child sexual abuse accommodation syndrome. I am confident that the patterns would continue to hold. Legal standards would be correlated with outcomes. Democrats and Republicans would behave differently and perhaps predictably. The state would win most of the time. And third-party validation would prove a tipping point if it were tailored to address legal questions. This is the true test of theoretical understandings. Returning to figures 2.1 through 2.2 on pages 31 and 32, it is possible to add the particular manifestations of key factors in the diagrams. Even while attitudes may be difficult to pinpoint, the location of the legal threshold and the amount of institutional political support could be used to calibrate the diagram and predict the result. If the legal standard was *Frye*, we would place the starting point further from the threshold. If the state rather than the defendant was arguing for admissibility, and/or there was a favorable third-party report present in the decision-making environment, we could increase the amount of institutional/organizational support. Given the above empirical work, it is possible to say that we have partially solved the puzzle of differential treatment and that we have discovered some reasonable expectations and predictions about the judicial processing of novel scientific evidence.

Important Political Implications

The policy implications of our understanding of gatekeeping decisions are the most important part of this analysis. What do these findings imply for the American legal system? First, legal standards work. Standards that control outcomes allow policy makers to calibrate the rate of judicial acceptance to the desired level. Legal standards are a source of real power in the judicial system. They are also a source of stability. The legal context (facts) is also important. With controversial, unsettled policy questions, such as with polygraph or syndrome evidence, judges were reluctant to foreclose all uses of the science or to open the floodgates. They were able to be responsive and adopt compromise positions to maximize the use of the science within the limits of unknown levels of validity and reliability. This ability of the legal system to adapt and evolve is significant for the effective processing of defendants. It also results in optimal use of the information provided by science.

Second, attitudes matter. The correlation between political attitudes and judicial outcomes suggests that the political system works—even with regard to science. Mass preferences can "trickle up" into judicial decisions. To the extent that judges share ideological attitudes with those within their jurisdictions, scientific jurisprudence will reflect those policy goals—even if regions only represent constraints on judicial attitudes. With regard to particular attitudes, this analysis found that the attitudinal dimensions at work for novel scientific admissibility in criminal courts had little to do with attitudes toward innovation and progress. Judicial sympathy with prosecutors and defendants suggests that attitudes were working to frame the problem as one of law-and-order policy orientation. In DNA and polygraph cases, Republicans consistently sided with the state at higher rates than Democrats. In the syndrome cases, Republican support for RTS was again synonymous with supporting the state. The usually high support by Republicans for BWS may be explained by (1) the fact that RTS had set the precedent, and BWS could not be readily distinguished; (2) the fact that defendants in BWS cases might have more sympathy for their crime; and (3) the fact that BWS cases are much more rare than rape cases in American criminal justice, and justices may not desire to jeopardize RTS jurisprudence with BWS dissents. Overall, the role of attitudes in scientific jurisprudence demonstrates that policy makers are aware of political winners and losers when it comes to novel scientific evidence. Decisions and cognitive processing are clearly affected by attitudinal framing and evaluation—allowing that politics, not science, may determine the outcome.

Third, science interacts with the judicial system vis-à-vis its relationship with law enforcement. Law enforcement consistently succeeded in obtaining favorable decisions from the judicial system. This is not necessarily a cause for alarm when we consider the importance of law enforcement in a civil society. It does mean, however, that science that is consistent with law enforcement goals is more likely to be developed and used in the legal system. If we consider law enforcement an arm of the government, and if the government is the institution commissioning authoritative third-party reports, the availability of third-party validation is also dependent on the power of the state. It is interesting to note that there was no NRC report on polygraphs. The main reports arose from the congressional Office of Technology and the Department of Defense, and they did little to settle the judicial debate.

Fourth, judicial policy as created by courts is largely left to "insiders." The role of policy advocates was very small. An analysis of amicus briefs found that legal personnel were by and large the only political actors providing additional information to the courts. This appears to contradict practices in other policy areas, such as civil liberties or contract law. The practice supports the assumption that scientific debates have not become salient to broader political interests. Furthermore, as alluded to above, legislatures may have usurped judicial power in this area. While civil liberties policy is ultimately settled in court, scientific admissibility can be decided by statute. The role of the legislature in gatekeeping policy will be revisited in the next chapter.

Fifth, the power to commission third-party validation is the power to open or close the gates on scientific evidence. Third-party reports are persuasive to gatekeepers. This has a positive side and a negative side. The power of third-party validation is positive for public policy because it is evidence that the judicial system is responsive to experts. This power is also problematic for public policy because experts may be aligned with government and law enforcement interests. We can imagine the difficulty of a diffuse, unorganized group of defendants to dedicate or marshal the resources required to obtain a report with the same legitimacy. Furthermore, judicial gatekeeping requires a partnership between legal experts (judges) and scientific experts to determine if a *way of knowing* is sufficiently valid and reliable for *legal* purposes.

Finally, this comparative analysis of state supreme court gatekeeping decisions demonstrates the ability of the federal system to process important policy questions. With a difficult policy question acting on the criminal system, the ability of state supreme courts to develop experimental approaches, arguments, and policies allowed the system as a whole to sort novel scientific evidence in an optimal manner. Imagine the way that gatekeeping jurisprudence would have developed if only one jurisdiction controlled the outcome. The possibility of less-than-optimal solutions would be more likely in a single-jurisdiction system. The federal system, however, with fifty autonomous jurisdictions is a very useful way to process complex policy questions. Eventually the most viable solutions surface—even if that solution is a stalemate (as with polygraph evidence).

Black Robes, White Coats, and the Politics of Knowledge

Legal commentators often seem exasperated with a judicial system that elevates soft sciences above the harder sciences. They want judges to act like scientists and make decisions according to scientific value systems. As outsiders, they idolize the lab and desire to bring its perceived precision and purity to the courtroom. They criticize the bench for allowing eyewitness testimony and fingerprint evidence while rejecting solid scientific findings, claiming that the former are much less reliable than the latter. Among legal commentators, there are constant calls for judges to exchange their black robes for white coats when dealing with scientific knowledge. Indeed, the federal government recently funded the Einstein Institute for Science, Health, and Courts (EINSHAC), an institute designed to provide judges with a crash course in scientific developments. The Department of Energy established the EINSHAC to research the role of science in trial courts and offer training conferences to judges who desire to improve their knowledge of scientific matters. There have also been calls for letting scientists, who have no political knowledge or jurisprudential understanding, determine public policy. According to a 1993 recommendation by the Carnegie Commission on Science, Technology, and Government for a Changing World, there is a need for special science centers to serve the judiciary: "To improve the quality of scientific and technical information that enters the courtroom and enhance the capacity of judge and jurors to evaluate and apply it in a legal setting, resource centers should be established within both the scientific community and federal and state judiciaries, and a nongovernmental Science and Judicial Council should be established to monitor and initiate changes that may have an impact on the capacity of courts to manage and adjudicate cases involving science and technology."

The legal commentators, at best, have an idealized view of how science works. The truth of the matter is that the judicial method and the scientific method necessarily approach knowledge differently. In their book *Judging Science* (1997), Kenneth Foster and Peter Huber describe this difference in terms of evidentiary standards. Scientists will (theoretically) use any and all relevant evidence in their investigation with no sense of admissible or inadmissible evidence. Law, on the other hand, necessarily requires explicit rules and evidentiary standards because of the value of justice. The goal of the judicial system is not necessarily truth. Rather, it is justice. Science does not worry about justice, at least not in theory. While Foster and Huber make a valuable distinction, they too have an idealized view of science. Science is also in the business of justifying. It is also guilty of foreclosing certain ways of knowing while pursuing others. In his book *Politics in the Laboratory* (2004), Ira Carmen effectively describes the role of politics in human genomic research. Furthermore, he makes the extraordinary claim that the study of politics can assist and direct the pursuit of scientific knowledge. This view correctly recognizes the role of politics in the processing of knowledge by both science *and* the judicial system.

An understanding of politics is essential to explain and predict judicial processing of scientific knowledge. Furthermore, an understanding of politics is likewise essential to explain and predict science's own pursuit of knowledge. In her book *Science at the Bar* (1995), Sheila Jasanoff outlines the relationship between law and science in a manner completely contrary to the manner desired by the legal commentators. Rather than science legitimizing the law, law has the effect of legitimizing science. According to Jasanoff:

> Science emerges . . . not as an independent, self-regulating producer of truths about the natural world, but as a dynamic social institution, fully engaged with other mechanisms for creating social and epistemological order in societies. . . . [T]he interplay of law and science acquires particular significance. . . . [L]egal disputes around scientific "facts" often appear as sites where society is busy constructing its ideas about what constitutes legitimate knowledge, who is entitled to speak for nature, and how much deference science should command in relation to other modes of knowing. . . . In each context, legal proceedings function as a medium for constructing and stabilizing particular orderings of science and technology in society. (42)

In this context, the study of gatekeeping decisions takes on new meaning; it is the study of a phenomenon with important ramifications for both political and scientific development. While the gatekeepers must take cues from other societal sectors, those societal sectors also take their cues from the gatekeepers. The laws and doctrines, the political attitudes, and the institutional and organizational leverage of the political game are as important to the evolution of science and technology as they are to the distribution of power and the politics of gatekeeping. By virtue of the interplay of this relationship, judicial processing of scientific knowledge allocates political values and resources for both the immediate jurisprudential players and society as a whole. Gatekeepers perform an important function in American politics and in the development of science and technology. Variation in judicial allocation of values and resources is a reflection of variation in the pertinent political processes.

CHAPTER 7

NEW CLUES?

GATEKEEPING AND THE TWENTY-FIRST CENTURY

Gatekeeping as Public Policy

If we think of judicial gatekeeping outcomes at the state supreme court level as gatekeeping policy for the jurisdiction, we can step back from the narrow question of admissibility to the broader question of policy outcomes. How do we predict policy outcomes? This book adopted the perspective of judicial cognitive processes and the immediate case factors that may inform those processes. This is, however, only one starting point.

The larger political-science question of predicting policy outcomes might also inform our analysis from several vantage points. If we are interested in the development of judicial policy or policy more generally, we can broaden our question and see how general theories of policy outcomes inform our expectations. This final chapter will move the discussion into additional realms of political science expectation. There are several broad theories of the policy process that may also do much to explain and predict judicial gatekeeping outcomes. There are also narrow expectations about additional particularized variables that might augment the above list. In the end, the idea is to step outside of the model presented in chapter 2 and examine other potential determinants of judicial policy outcomes in the gatekeeping context.

Models of the Policy Process

When attempting to predict a policy outcome, there are essentially four main schools of thought in the political science literature (Sabatier 1991): open systems theory, institutional rational choice theory, policy streams theory, and the advocacy coalition framework. Each of these systems would explain and predict policy outcomes differently, and they may have significant insights for understanding judicial gatekeeping decisions.

Open systems theory (Hofferbert 1974) was developed to explain differential policy decisions across different states. Hofferbert explained policy outputs as a function of history/geography, socioeconomics, mass political behavior, government institutions, and elite behavior. These variables are visualized as a broad "funnel" (history/geography) narrowing to immediate political actions (elite behavior). In the context of judicial gatekeeping outputs, particular state history and government institutions might play an important role in the development and use of particular scientific evidence. State history is an important part of state political culture. The research in this book uncovered a regional difference in several gatekeeping contexts, which might have something more generally to do with historical political culture development. (It could also have something to do with public opinion, ideology, or other variables that vary by state and might also be rooted in historical differences.) Socioeconomic position, as a measure of state wealth, might also explain judicial gatekeeping policy. It could predict differences in judicial professionalization and training, as well as important differences in the budgets available for law enforcement and criminal defense—variables closely tied to the "resources" measured as institutional and organizational advantages in the analysis of this book. It might be interesting, however, to see if budget differences among different judicial districts (particularly prosecutor budgets or lab infrastructure allocations) are closely tied to gatekeeping policy decisions.

The next factor in the "funnel" is mass political behavior. Theories about mass political behavior are less likely to factor into a gatekeeping decision simply because of its low public salience. Of special importance for the puzzle is the fact that salience is not likely to vary among publics in different states when it comes to scientific admissibility. One way that mass political behavior can be related to gatekeeping policy is in its relationship to the election cycle and the way that "law and order" politics might be influencing state politics at any particular time. This is especially likely to have an effect if gatekeeping decisions are being made in the legislative branch (see the legislative appropriation discussion below). Another way that mass political influence might explain variation in gatekeeping policy is if public awareness is particularly heightened in a particular state. This could occur because of a high-profile case. It might also occur as an interaction between the timing of a gatekeeping decision and the development of a cultural factor (such as the increased popularity of *CSI*, discussed below).

Government institutions could be correlated in a variety of ways, particularly in terms of judicial selection procedures, as well as the high court structure within the state. A few states have courts of criminal appeals that are the courts of last resort for criminal cases. The institutional difference between a high court specializing in criminal appeals and a high court with general appeal jurisdiction might be significant in the development of admissibility policy. In terms of elite behavior, the current analysis focused on judges. There is a significant possibility, however, that policy outcomes are produced by other elites as well—particularly the state attorney general. Of course the most obvious institutional effect is federalism (discussed separately below). The development of jurisprudence is largely a product of fifty state jurisdictions and thirteen federal court appellate jurisdictions. Gatekeeping policy in any jurisdiction may certainly be a function of the policy environment created by prior state and federal decisions.

As noted by Sabatier (1991), these variables are helpful in comparative analysis yet lack in theoretical understandings of how broad variables exactly work to form specific policy (the "black box" problem). The model of this book, as outlined in chapter 2, starts closer to the policy decision, with an emphasis on the decision maker and the forces shaping the proximate cues provided to the decision maker in a particular case. The value of a broader theoretical starting point is that it may find systemic variables responsible for patterns in policy outcomes. The regional variation is one example of this pattern, though the explanatory link requires additional work. Why is the Northeast or the West or the South treating questions of scientific admissibility differently?

A second theoretical approach, institutional rational choice (Kiser and Ostrom 1982), explains policy outcomes as the interaction of individual incentives with institutional structures. As mentioned above, judicial selection rules might be correlated with admissibility policy. This theory would suppose that elected judges and appointed judges have different goals and concerns, which are reflected in their policy outputs. While not obvious at first, judicial gatekeeping decisions might have an important relationship with judicial selection. In jurisdictions where judges are elected, especially in partisan districts where judicial races take on many of the trapping of legislative and executive races, judicial gatekeeping decisions can matter to the extent they are tied to electoral politics—most notably the need to court the favor of the bar, law enforcement, or the prosecutors association. In this situation, it is possible to model a criminal judiciary sensitive to law enforcement and prosecutorial needs simply because of the tremendous leverage of their political support during a judicial race.

The institutional rational choice model predicts that individual behavior will vary according to institutional rules and that the same individual would behave differently under a different institutional arrangement. It may be that those states with partisan judicial elections are also more favorable to law-and-order politics, which would imply a gatekeeping policy consistent with law enforcement goals. By contrast, states with appointed or tenured judges might exhibit a different set of incentives. This would allow those courts to favor interests generally underrepresented in judicial election politics, namely defendants. These hypotheses imply that differential treatment of science among jurisdictions might be correlated with judicial selection and retention methods.

Doctrine is another institution of interest from the point of view of institutional rational choice theory. Doctrines are institutions, and they vary from jurisdiction to jurisdiction. This is partly accounted for in the model from chapter 2. As an institutional rational choice component, however, it would hypothesize that individual behavior (among prosecutors and defense attorneys) and organizational behavior (perhaps in the larger supporting apparatus, such as commissioning research to demonstrate error rates) will vary whether the judicial standard is general acceptance or known rates of error. When an attorney prepares for a particular evidentiary standard, that attorney is responding to the current institutional structure. In the big picture, judicial gatekeeping policy can depend on whether a jurisdiction requires a *Frye* hearing. Evidence that might otherwise be deemed relevant may not have the scientific muster to pass a *Frye* hearing. These doctrinal institutions could also affect individual behavior at other levels. Perhaps prosecutors may seek to change the doctrine (as in advocating for the adoption of a *Daubert* standard) or they may seek to do away with the *Frye* hearing by lobbying for the legislature to declare certain scientific evidence statutorily admissible. In this manner, the institution and individual incentives would be responsible for the general gatekeeping policy.

Institutional rational choice also has a particular interest in the way that institutions at one level will inform another level of policy making. This is modeled at three levels: constitution, collective choice, and operational choice (Sabatier 1991). In gatekeeping, the institutions set in place by the U.S. Constitution and its jurisprudence lead to a special emphasis on death penalty habeas corpus appeals—in which scientific admissibility is often raised as an issue for state supreme courts. At the next level, the collective choice level, legislation such as public defender funding and automatic appeal laws is responsible for the routine appeal of these cases—leading to a constant barrage of scientific admissibility issues arising in state supreme courts. This might also account for variance in admissibility jurisprudence from science to science. Science associated with capital crimes will receive a different level of government work and a different amount of government scrutiny than science associated with white-collar crimes (e.g., forensic accounting). Scientific evidence associated with death penalty cases is expected to receive greater judicial attention, thus mobilizing additional interests with a stake in the outcome.

The operational choice level refers in part to the model in chapter 2, where policy is operationalized by a court and a policy is made for criminal prosecution in a particular state jurisdiction. Institutional rational choice forces the recognition of the role of higher order institutional work in the current distribution of outcomes—and it certainly adds explanatory power for gatekeeping policy in, say, criminal courts, versus civil litigation, where constitutional considerations are vastly different. Likewise helpful is the way that the institutional rational choice model incorporates feedback loops. Policy arising from state supreme court cases might feed back to the collective choice level. This is most obvious when the legislature usurps judicial power and writes admissibility policy themselves. This was a very real occurrence in the instance of rape trauma syndrome. Prosecutors who had lost at the operational level put pressure on the state legislative level to provide for statutory admissibility of rape trauma syndrome.

Stepping back from a focus on the individual is a systemic theory, the policy streams approach (Cohen et al. 1972, Klingdon 1984). This model, often referred to as the "garbage can model," returns to a broad picture of policy development. Rather than a funnel, this theory sees policy outcomes as the result of three separate "streams": the problem stream (events or developments highlighting the problem), the policy stream (people who have developed a solution), and the political stream (elections and legislative contexts that make it likely that the policy will be adopted). According to this model, policy is made when a window of opportunity is created by the convergence of these streams. This is fairly easy to visualize with gatekeeping decisions. Forensic DNA was developed as a technology at the same time that crime was on the rise. Reducing violent crime is likewise a popular election tool for local, state, and national officials. A very good example of this model with regard to DNA database policy (which implies gatekeeping policy) is a recent study by Neil Gerlach (2004) on the way that political streams, technological streams, and problem streams converged to produce a favorable policy environment. While Gerlach's research is couched in sociological theory, the underlying assumptions mirror the garbage can model. The policy streams model might also explain fluctuations in polygraph policy, partly because polygraph was the solution to witness veracity but posed a problem for legal doctrines that viewed fact-finding as a task given to the jury. This approach anticipates the success of an accommodating solution, such as stipulated results, and the window of opportunity that can encourage certain outcomes.

The policy streams approach may also explain why some states responded one way while others did not. Because a policy streams approach assumes that there are windows of opportunity, one can get the idea that placement on the timeline matters. The timing of a particular jurisdiction's decision becomes explanatory. The role of third-party reports and DNA admissibility jurisprudence in the research of this book illustrates this principle. The NRC officially declared DNA evidence an acceptable solution to the problem of perpetrator identification. Also, within each jurisdiction, it might be useful to consider the trajectory of gatekeeping policy for that state as an explanation for the presence or lack of a window of opportunity. It may be that certain states were more ready in terms of all three streams coming together at once.

A final model of the policy process emphasizes the role of advocacy coalitions (Sabatier 1988). Advocacy coalitions consist of actors who share a particular set of values (say law enforcement). They will work to change the rules to accommodate those goals. This certainly might be useful to explain the broader workings of the FBI, the president, and prosecutors/attorneys general to make forensic DNA legally viable in state criminal prosecution (see below). As visualized by Sabatier, the coalitions support model demonstrates the way that law enforcement coalitions are more successful than defendant coalitions. This approach places more emphasis on the role of policy advocates in gatekeeping policy outcomes. This model also incorporates the constraints imposed externally on coalitions and policy makers. In terms of gatekeeping policy, there is much to be said for the way that precedent can be a very strong constraint. This was obviously at work in early polygraph cases.

Each of these policy models offers additional insights for competing explanations of judicial outcomes. Additional empirical work would be necessary to see which models are best at understanding the work of the black robes in accepting the knowledge of the white coats. Each provides additional vantage points for the consideration of gatekeeping policy development in state jurisdictions, and the triangulation of several models significantly tightens explanation and prediction of gatekeeping policy.

Gatekeeping as Legislative Appropriation of Judicial Decision Making

The approach undertaken by this book examined only judicial behavior, with legislative behavior taken as exogenous to gatekeeping policy. A complete picture of gatekeeping policy, however, must recognize that not all gatekeeping policy is made by judges. State legislatures have done significant work in this area, and state rules of evidence have served to codify the admissibility of particular forms of scientific evidence.

By way of example, eleven states had codified DNA admissibility by 1999 (National Conference of State Legislatures), ten years after the first American DNA case. For instance, the Alaska code (§09.25.051) states, "Evidence of a DNA profile is admissible to prove or disprove relevant facts if the evidence is scientifically valid." Further examples can be found in Idaho (§19–5505, "DNA profiles may be used at trial as evidence provided the evidence is otherwise admissible at trial"), Maryland (§10–915, "In any criminal proceeding, the evidence of a DNA profile is admissible to prove or disprove the identity of any person, if the party seeking to introduce the evidence notifies the other party at least 45 days in advance and reproductions of the test and information must be provided to the other party if requested"), and Tennessee (§24–7–117, "Results of DNA analysis are admissible without antecedent expert testimony to the reliability of such evidence as long as the testimony meets the standards set forth in the Tennessee Rules of Evidence").

The dynamics of this "legislative appropriation" of judicial decision making are ripe for additional research on judicial gatekeeping behavior. The conditions under which individual state legislatures develop and implement policy in this area would inform several theoretical conclusions of this book. Theories of legislative policy making might make several contributions to a broader understanding of judicial gatekeeping. For instance, judges may or may not know what could trigger legislative appropriation of a judicial decision. If it appears that these decisions are being driven by powerful interests (prosecutors?) who have been thwarted in judicial outcomes, the judges may anticipate these maneuvers. This would imply that judicial gatekeeping policy favoring prosecution interests is the product of a conscious effort by judges to keep policy making under their control by placating prosecutors and keeping them from running to the legislature to "correct" judicial policy.

More empirical research is needed to know if legislative policy making regarding gatekeeping is in direct response to judicial outcomes. For instance, if the majority of state laws were passed to counter prior rejection decisions by a state supreme court, we could assume that gatekeeping policy is triggered by judicial decisions that thwart law enforcement. There is a need, however, to analyze the timing of gatekeeping legislation and judicial gatekeeping decisions. Some decisions considered in the research for this book involved judicial interpretation of gatekeeping legislation. The give-and-take of legislative and judicial gatekeeping considerations could be part of a larger story of competing interests (prosecutors and defense attorneys) using different institutions to protect their interests. Empirical timelines on the timing of legislation and judicial decisions is certainly an important part of understanding variation in judicial gatekeeping decisions and the broader sharing of policy-making power between the legislative and judicial branches. And while it was not considered for this book, it must be considered in future discussions of judicial policy making with regard to scientific evidence.

Of similar concern is the genesis of legislative policy making. Legislative gatekeeping policy (or its absence) could be determined by a host of significant factors. Perhaps a policy entrepreneur (such as the attorney general or a leading legislator) is pushing for codified gatekeeping policy. It might also be that the larger political context, such as an election year, is motivating state legislative leadership to try to please both prosecutors and women's rights advocates by voting to codify rape trauma syndrome admissibility. Legislative gatekeeping policy can also be the result of certain motivating events, such as a high-profile trial involving battered woman syndrome, where a gatekeeping issue is put on the agenda. Each of these examples has positive implications for the ability of the democratic process to implement the larger political will.

Yet, gatekeeping decisions appear to be of low public salience—especially in terms of judicial policy outcomes, such as the ones analyzed in this book. As the data on policy advocates demonstrate, most amici briefs were from judicial insiders. Is it possible, then, that legislative gatekeeping policy might also be the product of insider activity, with little influence from the general public. Is this more akin to technical bureaucratic work? A study of the legislative development of gatekeeping policy via legislative records of sponsors, hearings, committee work, and debate might serve to demonstrate that a handful of people were pushing the change, with little conscious consideration of larger political gains (such as electoral outcomes).

Also of importance is the fact that most legislative gatekeeping activity is admitting evidence. Only in the case of polygraphs do we see legislation limiting admissibility by statute. There are some interesting political stories embedded in an institution that is routinely used to admit (rather than reject) evidence. At the most basic level, it is likely the case that law enforcement and prosecutor interests are active at the legislative level because most scientific evidence is useful for the prosecution. It may also be that prosecutors are using this method as a way around unfavorable judicial decisions. It might also be that admissibility is embedded in larger legislative activities, such as budgeting for law enforcement training and state laboratory infrastructure.

Also of interest is the extent to which the factors of judicial policy central to this book, such as partisanship and third-party reports, affect legislative gatekeeping policy. If Republican judges are ruling in the same direction as the votes of Republican legislatures, there may be a larger partisanship issue attached to gatekeeping policy. Likewise, a partisan component might be present if Democratic judicial behavior mirrors Democratic legislative behavior with regard to gatekeeping policy. It might be, however, that judicial and legislative partisanship do not line up. This would serve to test which political attitudes are informing policy decisions. As noted in chapter 2, judges of the same party may have different reasons for voting in the same direction. Some may view the decision as one about law-and-order politics, while another sees it as a policy innovation. The relationship between partisanship could be further tested with consideration of legislative support for gatekeeping policy.

The same is true for third-party reports. First, were they a factor in the legislative work? Do they appear in the legislative record? Were they used in the same manner as the judicial branch used them? Or do legislatures emphasize different portions of the same report? Furthermore, is there any evidence that the legislature commissions its own report, such as the Office of Technology Assessment's report of polygraphs, which was undertaken at the request of Congress? The ability of legislatures to commission reports is a part of the policy story that differs greatly from the adversarial judicial context, where courts cannot and do not generate independent sources of information about a particular decision.

A separate factor of judicial outcomes, legal doctrine, and the evidentiary standard at work in the district also has important legislative implications. At the most basic level, do legislatures in *Frye* districts find themselves more active in gatekeeping legislation than legislatures in *Daubert* districts? If *Frye* districts make it harder to admit evidence (as was implied by the evidence in this book), it may be that thwarted interests in those districts are quicker to pursue a legislative solution for their gatekeeping goals. It is also of interest to see if the legal standards themselves have been codified. It may be that changes in science and technology encourage legislatures to police judicial activity by dictating the admissibility standards to be used in the jurisdiction.

In general, attention to legislative gatekeeping activity is very important for a complete view of gatekeeping policy. Codification of gatekeeping admissibility has immediate and important implications for winners and losers in the criminal justice system. Also not to be neglected with regard to legislative activity is the power of the judiciary to interpret admissibility statutes. The statutes excerpted above leave quite a bit of room for the judiciary to continue to have the final word in individual gatekeeping cases. Furthermore, it is possible that the judiciary can modify or cabin legislative intentions using its role as interpreter of the law. Thus additional research into judicial policy outcomes must view them in the context of actual or potential legislative activity.

Legislative support can arrive in other forms than admissibility statutes. A recent congressional act, the Equal Justice for Women in the Courts Act of 1994 (SEC 40401), provided grant money to states to, among other things, train judges and court personnel in "the use of expert witness testimony on rape trauma syndrome, child sexual abuse accommodation syndrome, post-traumatic stress syndrome, and similar issues" (SEC 40411 § 8). Other judicial education initiatives, such as the Einstein Institute for Science, Health, and the Courts (EINSHAC, sponsored by the Department of Energy), are likewise directly aimed at educating judges on scientific developments for the purpose of encouraging the use of science in judicial proceedings. EINSHAC, recently renamed Advanced Science and Technology Adjudication Resource Center (ASTAR), has initiated a series of conferences for judges to discuss controversies surrounding molecular biology. It was created by a congressional mandate to prepare judicial personnel for the impact of the Human Genome Project. In 2007 alone, 39 jurisdictions and 214 judges attended the National Judges' Science School (Advanced Science and Technology Adjudication Resource Center 2007).

In terms of our gatekeeping puzzle, these realities suggest a possible role for pinpointed judicial education as a predictor of judicial behavior. Judges receiving this training could theoretically create gatekeeping policy that is qualitatively different from judges receiving no special training. These programs certainly assume a behavioral difference in attendees and non-attendees.

As the above discussion illuminates, legislative activity is important for gatekeeping policy. Legislative activity can codify gatekeeping decisions, with an obvious bias toward admissibility. Partisanship and third-party reports might also affect legislative outputs and gatekeeping policy more generally. With regard to judicial decision making, legislative policy can be a function of the legal admissibility standard at work in the jurisdiction. Also of importance is the way that judicial activity may be shaped by interpreting legislative admissibility statutes. Finally, legislative activity can affect judicial outcomes by creating judicial education opportunities. Thus, the model of state gatekeeping variation in this book is only part of the potential story of factors influencing gatekeeping decisions.

Gatekeeping as Executive Branch Leadership

The legislature is not the only area of concern when it comes to understanding judicial gatekeeping policy. An important additional clue to gatekeeping behavior must be lodged in the executive branch. Gatekeeping policy initiatives can arise at several important levels in the executive branch. At the bureau level, there may be important initiatives undertaken by law enforcement or prosecution policy. For instance, the FBI initiated vigorous initiatives to gain state judicial acceptance of forensic DNA, such as the Violent Criminal Control and Law Enforcement Act of 1994 (implementing uniform standards for DNA testing). (No such initiative was undertaken for syndrome evidence.) It has published law enforcement resources, such as the 1999 brochure entitled *What Every Law Enforcement Officer Should Know About DNA Evidence* (National Institute of Justice 1999), a pocket-sized brochure providing practical information about identifying, preserving, and collecting DNA to help solve cases. The National Criminal Justice Reference Service also spotlights prosecutor resources, such as *Forensic DNA Fundamentals for the Prosecutor—Be Not Afraid* (American Prosecutors Research Institute 2003) and *DNA Evidence Policy Considerations for the Prosecutor* (APRI 2004). Other prosecutor resources include publications with tips for enhancing juror understanding of DNA evidence, such as *Can Jury Trial Innovations Improve Juror Understanding of DNA Evidence?* (Cann, Hans, and Kaye 2006). FBI resources even target lawyer and judicial education with the development of online DNA training (Schmitt 2007).

The FBI also worked deliberately for DNA admissibility in state court jurisdictions. Early on in DNA legal battles, the FBI was confident about overcoming legal challenges to DNA admissibility (Hart 1998). In terms of directly affecting gatekeeping policy, the FBI sponsored the NRC report in 1992 and worked vigorously to obtain a completely supportive endorsement by 1996 (Neufeld 1993). According to Neufeld, the assistant director of the FBI was personally involved in shaping the NRC reports. This kind of direct effect is important when we consider that these reports essentially silenced the debate on scientific validity.

Other examples of FBI support for scientific evidence are likewise quickly forthcoming. For example, in the face of mounting unfavorable scientific evidence, the FBI has continued to vigorously support polygraph use in criminal and employer settings. With so many critics of the science, the continued use of polygraph evidence is significantly explained by FBI emphasis on its utility—even in the face of contrary scientific reports (Warner 2005). The Department of Justice is likewise active in supporting scientific evidence more generally. They have various education programs pertaining to scientific evidence, such as the National Victim Assistance Academy, which has prepared education materials for recognizing victim syndromes.

Also of interest with regard to explaining admissibility policy would be the role of executive actions taken on behalf of a particular science in criminal prosecution. The most grand example of this would be the president's DNA initiative (Advancing Justice Through DNA Technology 2003), a multipronged approach to using DNA to the fullest in criminal prosecution. The initiative includes education, training, and infrastructure for DNA data banks and DNA typing. The federal government maintains an official Web site (http://www.dna.gov) for all things pertaining to law enforcement, prosecution, and DNA evidence. This initiative will certainly be responsible for subsequent DNA admissibility issues, particularly with the proliferation of DNA data-bank maintenance.

Presidents also appoint task forces to make policy recommendations with regard to the role of science in criminal prosecution—especially when it comes to finding ways that law enforcement and prosecution can more effectively use scientific advances. For example, rape trauma syndrome figured prominently into the President's National Advisory Committee for Violence Against Women, which has released several reports in recent years (http://www.usdoj.gov/ovw/nac/welcome.html).

Executive roles can take on other meaningful theoretical relationships to judicial policy outputs. First, the executive use of a scientific method (such as the pervasive use of polygraphs in the Department of Justice and the Department of Defense) can affect judicial outcomes. Use of a science implies implicit approval of its reliability and validity, and it is difficult for a court or attorney to argue that a method is unreliable or invalid when it is viewed as a legitimate tool in other government work. Indeed, many of the cases in the polygraph chapter involved judges (either in the dissent or the majority) arguing that polygraphs were accepted as legitimate for other venues.

Second, executive lobbying for science at the state level can significantly affect state admissibility rates. Virginia, by virtue of its proximity to Washington, DC, was the first state to admit forensic DNA evidence. It may be that admissibility cases can be explained by concerted federal efforts in a particular jurisdiction. This effort could occur as part of law enforcement education, training, and best practices. It could also occur as consulting or amici work for particular cases. It could also be authentic lobbying in the sense of working at the level of the state legislature to change or amend admissibility statutes.

A variant of executive lobbying is the work of particular policy entrepreneurs within executive organizations. For example, a speech by an FBI director might provide political authority (or lack thereof) for a particular type of science. The most important policy entrepreneur, the president, can certainly affect scientific admissibility. According to a White House publication entitled *Advancing Justice Through DNA Technology* (2003), the president's DNA initiative promoted (1) using DNA to solve crimes, (2) using DNA to protect the innocent, and (3) using DNA to identify missing persons. As part of crime solving, President Bush's specific initiatives included several directives for the U.S. attorney general, including eliminating backlogs, strengthening crime lab capacity, stimulating research and development, and providing training for "police officers, prosecutors, defense attorneys, judges, forensic scientists, medical personnel, victim service providers, corrections officers, and probation and parole officers" (iii).

Like the legislature, executives have a significant role to play in judicial gatekeeping variance. The interaction of state case timing and executive branch activities can significantly affect the judicial policy environment. FBI programs and publications and presidential initiatives cannot be ignored as part of the story of judicial gatekeeping policy formation.

Judicial Influence

When it comes to explaining judicial outcomes, much can be said for the role of the judges themselves. While this research tested only partisanship and regional variables, many more explanatory variables exist. Chief among them are the role of judicial interest groups (such as the American Judges Association) and judicial training initiatives (such as the ASTAR Resource Judge Program's National Judges Science School).

A thorough analysis of judicial admissibility should look at the professional media and professional organizational literature targeted to judicial practitioners. Official positions, sponsored panels, and special journal editions all send signals to judicial personnel about the feelings of their professional association with regard to scientific evidentiary issues. Judges' journals and professional associations certainly take positions on these developments, and an examination of their literature, media, and conference topics might yield important findings of influence and insight. Of particular importance are specific conferences or professional initiatives. The timing of these in relation to specific judicial decisions might be correlated.

A second explanatory factor could be judicial training and education. This could be measured in terms of time (newer JDs are expected to be more science friendly than older JDs). This is especially easy to imagine when we consider that the discovery of DNA occurred in 1952. Judges whose formal education predates DNA's entrance into biological curriculum might certainly have a more skeptical view than judges familiar with DNA concepts. Of even greater importance for policy initiatives, such as the president's DNA initiative, is the assumption that judicial hesitancy is simply a matter of lack of education and training. There are important theoretical reasons to believe that judges who are well trained and well versed in science might reach significantly different conclusions regarding the validity and reliability of scientific evidence than nonexpert judges. A study by the National Council of Juvenile and Family Court Judges (Raloff 2005) found that judges lacked significant understanding of basic scientific principles, such as hypothesis testing and the significance of error rates. This is a real problem for judges living in a post-*Daubert* world in which the standard assumes a solid understanding of these benchmarks.

What is interesting is that both the president's initiative and commentators who call for "science courts" assume that scientific training will make judges more likely to admit scientific evidence. This is not a foregone conclusion. With the advent of *Daubert*, it is just as likely to imagine that a science-trained judge would be more skeptical of certain ways of knowing than the average judge. Scientific training may aid in promoting lab evidence while discouraging couch testimony. Thus, an additional clue to judicial behavior must certainly rest in the scientific expertise of a particular court or jurist. It might be useful to look at admissibility rates for judges who have (or have not) attended the National Judges School.

Also of interest is the way that judges view the role of science in judicial settings. Researchers have uncovered patterns of admissibility where judges refuse to accept science considered perfectly valid by scientific entities. For example, one researcher found that judges refused to admit "case studies" in half of a sample of toxic tort cases (Raloff 2005). Other scholars have classified judges as idealists or pragmatists based on their approach to science and scientific evidence (Caudill and LaRue 2006).

In terms of a broader consideration of judicial admissibility, it is important to remember that the judicial lens involves considerations beyond validity and reliability—also of interest are various legal considerations, which in and of themselves may independently affect admissibility of science. In a policy forum appearing in *Science*, legal experts reminded scientists of the additional legal requirements involved in judicial admissibility decisions (Hoffman and Rothenberg 2005). Requirements for civil procedure, Fourth Amendment considerations, and relevancy standards are each legal constraints on judicial behavior. These can be considered a larger version of the "legal standard" hypothesis advanced in the research presented in this book. The underlying premise remains the same: legal rules might be the best explanation for variation in admissibility. Analysis of additional legal standards, however, would also serve to explain gatekeeping policy more generally. It may be that a particular type of science has trouble meeting the relevancy standard, or that a particular state has unusual requirements for civil procedure.

Gatekeeping and Federalism

Unique to American criminal justice and the development of judicial policy is the work of federalism. Federalism refers to the reality of fifty autonomous state jurisdictions (state supreme courts and state legislative codes) and thirteen autonomous federal jurisdictions (U.S. Circuit Courts of Appeal). Federalism is important because gatekeeping policy is being developed separately and sometimes simultaneously by different, yet equally situated, gatekeepers.

There are several important ways that federalism can affect gatekeeping variation. One effect is the role of diffusion and the likelihood of policy convergence among state and federal courts (Kilwein and Brisbin 1997). State courts may look to the policies of previous iterations of the gatekeeping question in other districts. This is quite obvious in many, if not most, of the judicial decisions considered in the analysis of this book. The judicial activity occurring in other jurisdictions prior to a current decision are certainly an important part of the gatekeeping story. Additional empirical work on this relationship is necessary. A starting point would be to look at the degree to which earlier decisions in other jurisdictions are cited—and the extent to which these prior positions are adopted by the court currently crafting its own policy. In a policy environment with conflicting prior decisions, it might also be possible to explain or predict the present court's actions as following one jurisdiction over another based on perceptions of court prestige. Caldeira (1985) developed a measure of state court reputation based on the number of times various state supreme courts were cited by their peers. This effect would be modeled as peer influence based on court reputation—and it is quite possible that judicial gatekeeping decisions are partly due to the emulation of prestigious peer courts.

General stories of policy diffusion among state court jurisdictions have also cited other factors associated with emulation of peer judicial policy, such as national events (Glick 1992). Scholars have even given state courts "innovativeness" scores (Canon and Baum 1981). It may be that certain courts are simply more likely to innovate with regard to judicial policy generally, and these courts gain a reputation as a policy leader, thereby increasing their reputation and likelihood of being emulated. In terms of gatekeeping policy, it might be important to compare emulation with reputation to understand why some gatekeeping decisions will be copied while others will not. Even more important, later decisions can be explained as "reinventions" of earlier decisions by other jurisdictions (Glick 1992). This might especially explain the diffusion of polygraph gatekeeping decisions in the 1970s.

Federal judicial decisions can certainly have an effect on state judicial decisions. By way of example, consider the 1995 decision *United States v. Galbreth*, 908 F.Supp. 877 (D. N.M. 1995) by a federal tax court to declare polygraph testimony valid and reliable under *Daubert* (Garmisa 1996). This could very likely affect additional jurisprudence in this area by signaling to lower courts that a higher courts finds polygraph evidence satisfies legal requirements for scientific admissibility. Additional analysis of state gatekeeping decisions could overlay important federal decisions on the timeline. A pattern of federal emulation or at the least influence is likely to occur. Scholars have noted that federal courts often lead the way in policy development (Tarr 1994; Haas 1981). It would be useful to chart a timeline of federal and state decisions to see if patterns emerge with regard to gatekeeping policy toward specific forms of scientific evidence. It would also be useful to see if state supreme courts are citing federal decisions to support their own conclusions.

Differential Treatment of Different Kinds of Science

A broad theory of gatekeeping behavior would combine the research of this book for the development of a higher-order research question. Rather than explaining the puzzle of state variation with regard to a particular science, such as hair analysis or child sexual abuse accommodation syndrome, it would seek to explain the puzzle of scientific variation: why is DNA less controversial than polygraph evidence? In other words, the gatekeeping phenomena can be attacked at a higher level of analysis: namely, asking the question, Why is one science rejected while another science is accepted?

This broader theory of scientific admissibility and gatekeeping behavior begins by imagining several possible gatekeeping patterns. First, you could have a situation where initial decisions about a particular science were in sharp disagreement. Some states are admitting while other states are rejecting. If this is the case, two scenarios are possible over time: convergence or continued divergence. Convergence refers to a pattern where jurisdictions reach a uniform policy about the admissibility (or rejection) of a particular way of knowing. Thus we could talk of science with a positive pattern of convergence and science with a negative pattern of convergence. Positive convergence occurs when judicial disagreement among jurisdictions converged in a positive direction over time, such as it did for DNA. A second pattern would be the reverse: negative convergence. While initially experiencing judicial disagreement, this pattern resolves the question of admissibility in a negative direction.

Divergence is the opposite of convergence. Here, the initial disagreement among states continues for the entire timeline. In terms of continued divergence, it is possible to imagine that the controversy may never be resolved, as seems to be the case with current rape trauma syndrome jurisprudence. Initial rejection/acceptance divergence continues. This kind of divergence might also be characterized as an indeterminate pattern, with jurisdictions continuing to fall on each side of the timeline in no discernible pattern.

There is an entirely different set of patterns arising from a situation where all initial decisions are in the same direction (all admitting or all rejecting) until a permanent shift occurs. This could be a positive permanent shift (where rejection decisions become admitting decisions). This could also be a negative shift (where what used to be admissible has now become rejected) as might be occurring with fingerprint evidence.

Third, it is possible to have a situation where initial decisions are in a single direction while subsequent decisions are divergent, with some favoring and some not favoring. This is similar to the picture of battered woman syndrome. Initial cases admitted BWS testimony while subsequent cases both admitted and rejected.

Fourth, it is possible to have a dynamic pattern, where initial decisions shift in a similar direction, only to shift back to the initial decision at a later date. This is certainly one way to describe polygraph admissibility patterns. Initial decisions to reject shifted to limited admissibility in the 1970s only to return to a dominant pattern of rejection in the 1980s.

Given these possible trajectories (converging, diverging, shifting, and dynamic), it might be possible to develop theories about the factors responsible for the trajectory pattern. Likely variables would have to include a political story. For instance, it might be that institutional changes (DNA data banks), doctrinal changes (easier or harder judicial standards), or changes in winners and losers (e.g., the use of DNA evidence in post-conviction relief) might be part of the story of evolving judicial policy. It might also be that particular shifts or dynamics would be tied to significant events or developments (such as the publication of a key report). When thinking about variation in trajectory patterns from science to science, it might also have something to do with larger political work, such as executive or legislative support, or the ascendance of political support forces within the executive or legislative branch, such as the rise of law-and-order politics. Trajectory variation would certainly be a function of the state of the art of the science itself. It may be that lab forensic science consistently demonstrates a positive convergence or a positive shift. It may be that psychological evidence is much more consistently divergent. These findings would say something about science, but also something about political support for particular types of science.

Gatekeeping and the Media

A very different host of variables arise when we consider the role of the media in public policy outcomes. The media can play several roles. For our purposes, we will consider two direct ways in which the media might affect judicial gatekeeping decisions.

One manner of media interference is in the nature of direct commentary on a particular case or issue currently before a court. This model of media interference assumes that the media can directly affect the outcome in a particular state by reporting on a particular gatekeeping case. Using media powers, such as framing or interpreting the story, the media may have an impact on judicial gatekeeping decisions. In order to create a robust theory of the role of media in this area, we would have to consider several variables: the magnitude of media involvement (for instance the number of pages committed in print to the particular case at issue). We would also have to consider the direction of media involvement: was it unified toward admissibility (rejection) or were there competing views represented? A third consideration would revolve around the types of media and hypotheses about how different types of media are likely to influence judicial decision makers. For instance, statewide attorney media, such as the *Minnesota Lawyer*, *Chicago Daily Law Bulletin*, or *Florida Bar News*, certainly report on gatekeeping decisions or developments. Consider the following examples from state legal news outlets:

- Brian Mackey, "High Court Urged to Press on DNA Testing," *Chicago Daily Law Bulletin*, August 24, 2006.
- Tony Anderson, "Wisconsin Supreme Court Considers DNA Testing Standards," *Wisconsin Law Journal*, April 27, 2005.
- Stephen P. Garmisa, "Chemical Syndrome Flunks Frye Test," *Chicago Daily Law Bulletin*, January 12, 2006.
- Barbara Jones, "Latest DNA Technology Fights for Acceptance in MN," *The Minnesota Lawyer*, May 27, 2002.
- David E. Frank, "Battered Women's Syndrome Gaining Acceptance with Courts," *Massachusetts Lawyers Weekly*, May 21, 2007.
- David Heckelman, "Justices Use Murder Case to Rule on Assailability of DNA Match Method," *Chicago Daily Law Bulletin*, November 12, 1996.

These articles are part of the media environment of the jurisdiction. When gatekeeping decisions are considered, any favorable (or unfavorable) media attention to the issue may have an effect on judicial outcomes. Thus, it might be possible to understand some gatekeeping decisions in light of media discussions occurring in the jurisdiction or with regard to the particular case or both.

Media outlets can also affect policy more generally by giving particular views a place in the policy environment. Especially important may be the way that media outlets affect practitioners. Media outlets such as *Judges Journal* or *Trial Advocacy* are written for particular members of the judicial policy-making context. Gatekeeping discussions occurring in these media outlets might also have an effect on judicial policy outcomes by shaping member values, expectations, or perceptions. Consider the following examples of gatekeeping articles in professional media:

- Donald A. Dripps, "Polygraph Evidence: Winds of Change?" *Trial* 33, no. 12 (December 1997): 77–76.
- David Gallai, "Polygraph Evidence in Federal Courts: Should it Be Admissible?" *American Criminal Law Review* 36, no. 1 (January 1999) 87–116.
- Kelly McMurry, "Judicial Debate over Repressed Memory Syndrome Continues," *Trial* 31, no. 8 (August 1995): 81–82.
- Lenore Walker, "Understanding Battered Woman Syndrome: Victims and Violence," *Trial* 31, no. 2 (February 1995): 30–36.
- William C. Thompson, "DNA Evidence in Criminal Law: New Developments," *Trial* 30, no. 8 (August 1994): 34–42.

These articles suggest that some gatekeeping policy is likely the result of direct media influence. As noted in chapter 2, judges are policy makers acting under uncertainty—they do not know if a particular scientific method is valid and reliable. They must rely on proximate sources of information. One of these sources of information is the state of the art in the profession. The perceived acceptance or rejection of a particular science as judicially worthy sends influential messages to judicial policy makers. To empirically test this effect, it might be necessary to compare subscriptions, reading habits, and judicial decisions. In other words, among those who received the publication, read it, and made policy in that area, was there a correlation between the direction of the article and the direction of judicial outcomes? Concordantly, is the correlation less for non-subscribers? Measuring direct media effects is extremely difficult, but this does not mean we should ignore them as a clue to judicial behavior—especially for policy involving significant uncertainty on behalf of the policy maker.

Gatekeeping and Popular Culture

The other effect of the media on scientific gatekeeping decisions is not due to direct reporting, but rather on the way that media shapes public and official views of science and the law by means of its communication and development of popular culture, beliefs, and perceptions. This cultural effect arises from fictional media depictions of science and the legal system: "T.V. has become our principle storyteller, transmitting legal norms or, arguably, creating them" (Cole and Dioso 2005, 13). This creation of public and judicial expectations in recent culture has been labeled the *CSI* Effect. The *CSI* Effect is a phenomenon of changing perceptions and expectations due to the popularity of the television program *CSI: Crime Scene Investigation* and its spin-offs (*CSI: Miami, CSI: New York,* and *Crossing Jordan*). A key element of the *CSI* Effect is that it can create unrealistic expectations for juries (Cole and Dioso-Villa 2007). Scholars have identified several possible outcomes of the *CSI* Effect, particularly in terms of expectations for evidence. Both prosecutors and defense attorneys are worried about the way that *CSI* may influence jury decisions. Prosecutors worry that juries will expect too much scientific evidence. Defense attorneys worry that juries will place too much weight on scientific evidence. *CSI* has also had an effect on juror technical knowledge, and there is a sense among prosecutors and defense attorneys that jurors are increasingly sophisticated about scientific evidence. Police chiefs are even worried that criminals will learn too much from TV crime-scene shows. There is also an increasing interest in forensic science education and training programs among college students.

CSI is not the only the only instance of TV cultural influence on science and the law. In the 1980s there was a television program called *Lie Detector.* In this reality show, guests were interviewed and given real polygraph tests by a licensed polygraph examiner ("Kansas Supreme Court Rules" 1984). This show and other media depictions of polygraphs, such as Fox's *Moment of Truth* (2008), can significantly affect perceptions of scientific reliability and validity. Certainly, the pervasive use of science in television programs could give the impression that these are routine and commonplace—and, by extension, reliable.

In terms of judicial gatekeeping decisions, it is possible that these cultural developments may have a trickle-down effect on judicial decision making, either in the form of direct education or in the form of perceived public opinion. For this reason, empirical work correlating TV programming on scientific evidence and judicial gatekeeping policy (or perhaps legislative gatekeeping policy) might be warranted. Because *CSI* has had such a large audience (twenty-five million viewers a week in 2005–2006 according to the Nielsen Media Research, Difonzo and Stern 2007), it is the show expected to have the largest effect on law and legal policy.

What does this imply for judges and scientific gatekeeping decisions? How might *CSI* affect judicial outcomes with regard to the admissibility of scientific evidence? Scholars have found judges referring to *CSI* in their opinions (Difonzo and Stern 2007). There is some worry that a lack of scientific education on behalf of judges may lead them to likewise be susceptible to technical scientific presentations. Scholars point to cases where judges have admitted dubious science, such as ear-print analysis, facial mapping, and an analysis of lip reading from video images (Difonzo and Stern 2007).

Recent research on jury effects has found little concrete evidence of differential jury verdicts due to *CSI* expectations (Cole and Dioso-Villa 2007). According to these scholars, jury simulation evidence provides little support for a *CSI* Effect on criminal prosecution. What they did find is that prosecutors may be altering their behavior based on a perceived *CSI* Effect among jurors. Prosecutors are assuming that jurors have higher expectations and technical sophistication, and they assume educational and persuasive roles with regard to these expectations. Prosecutors may include lectures about the difficulty of obtaining forensic evidence in the crime scene at issue. Or they may play up the strength of DNA evidence while neglecting conflicting eyewitness accounts or flimsy alibis.

This finding is particularly interesting for our purposes. It is possible that judges will make decisions based on their perceptions of juror behavior. In other words they could raise the bar for scientific admissibility if they felt that the jury was too easily persuaded by technical forensic evidence. Thus, the prosecutor study might mean that other personnel in the criminal justice system, such as judges, are also making adjustments to their policies based on a perceived effect. This data might indicate a change in law after the dissemination of the *CSI* Effect among judicial personnel. In terms of jurisdiction policy, rather than individual trial rulings, a *CSI* Effect is less likely. Here the legal standards and relevant institutional and organizational resources are much more likely to influence the outcome. There is little way to conceive of a *CSI* Effect for general gatekeeping decisions. It might also be, however, that *CSI* viewing itself might have an effect on individual judges, perhaps making them more willing to admit forensic evidence. It might also have an effect on legislators working on gatekeeping legislation.

Difonzo and Stern (2007) offer yet a different political use of a *CSI* Effect. They contend that it may be media created and perpetuated by prosecutors and defense attorneys using its perceived effects for their own advantage in the courtroom. In this context, *CSI* becomes an institutional/organizational advantage modeled in this book. Indeed, the interest in forensic science has served to increase government funding for crime labs (Cooley 2007). This implies that cultural entertainment might have an indirect effect on gatekeeping outcomes by privileging certain science (ways of knowing) above others, and thereby creating differential levels of political support and differential levels of resource allocation.

The effect of cultural beliefs about science more widely has been documented in judges. Caudill and LaRue (2006) have created a typology of judges based on content analysis of judicial opinions. According to Caudill and LaRue, judges can be sorted into two types: ideologues and pragmatists; they also note that many legal practitioners idealize science to the point of being too strict or too lenient when it comes to admissibility decisions.

New Clues?

Judicial gatekeeping policy is a function of law, institutions, organizational resources, and judicial attitudes. It is also a function of the interaction of the American political system, where multiple independent branches of government have a direct and measurable hand in gatekeeping policy development. As the discussion above has illustrated, legislative and executive action, as well as the action of peer courts, can significantly affect gatekeeping policy and gatekeeping jurisprudence. State variation may well be the product of these forces and the timing of judicial decisions in the larger political conversation occurring at multiple levels of government.

Judicial gatekeeping policy is also a function of the interaction of a particular science with political realities. The puzzle of variation in the treatment of science—one policy trajectory for DNA, another for battered woman syndrome—is a separate research question. Yet, understanding the trajectories of particular science/law interactions is an important part of the gatekeeping puzzle. If policy converges for some sciences and not for others, the political causes of these disparate paths must be understood. The research of this book leads one to hypothesize that pro-prosecution science is much more likely to experience convergence—except when the science significantly conflicts with institutional goals (such as jury fact-finding values and polygraph evidence). It is also possible to hypothesize that shifts in trajectories will occur with the destabilizing force of third-party reports (the 1992 and 1996 National Research Council reports on DNA) or legal mutation (the introduction of stipulated polygraph results).

Finally, judicial gatekeeping policy is subject to the forces acting on all policies in a democracy. Forces in the popular culture shaping the beliefs and preferences of citizens and policy makers have a very real effect on policy outcomes. With regard to gatekeeping, the effect of popular culture on scientific admissibility has been cultivated by fictional television programs about forensic evidence. Media influence, more generally, is also powerful and important, even when gatekeeping decisions have low public salience.

Conclusion

These additional clues to judicial behavior are a starting point for additional empirical work as well as an end point about the limits of the explanatory power of the current model. If we return to the walled-city metaphor in the beginning of this book, we can revisit with fresh eyes our model of the gatekeepers. As we watch the gatekeepers of our ancient city approving some visitors while turning away others, we begin to see important patterns. Some gatekeepers admit more than they reject. These are the gatekeepers with a significantly lower threshold for admissibility. Some gatekeepers consistently admit those in the company of city officials. These are gatekeeper subject to the organizational and institutional forces favoring law enforcement. Some gatekeepers are known among outsiders for their attitudes toward particular visitors. These are gatekeepers with particular persuasions (or from particular regions) known to favor some visitors over others.

If we were to change our vantage point and examine changes in gatekeeping policy over time, we might find diffusion patterns among the gatekeepers. We might also find that official edicts, support from local elders, new needs, or the spread of popular beliefs changed the admissibility patterns for particular groups of visitors. If we were to look at the experiences of one group of visitors—as compared to another group of visitors—we would find that they experience similar or different trajectories of approval or rejection. It may be that gatekeeping policy converges on the acceptance of a certain class of visitors while remaining far from uniform with regard to another class of visitors.

One reality becomes more and more apparent—there are patterns at work in this system. There are ways to explain and predict gatekeeping policy. There is a solution to the puzzle of differential treatment from gate to gate and from visitor to visitor. With enough observation and empirical data collection, it is possible to take much of the mystery out of the gatekeepers.

APPENDIX A

STATE SUPREME COURT CASES FOR FORENSIC DNA

State	Case	Citation	Year	Outcome
Alabama	Ex Parte Perry	586 So.2d 242	1991	0
Alabama	Ex Parte Hutcherson	677 So.2d 1205	1996	0
Alabama	Turner v. State	746 So.2d 355	1998	0
Alabama	Ex Parte Taylor	825 So.2d 769	2002	1
Arizona	State v. Bible	858 P.2d 1152	1993	0
Arizona	State v. Johnson	922 P.2d 294	1996	1
Arizona	State v. Hummert	933 P.2d 1187	1997	1
Arkansas	Prater v. State	420 S.W.2d 429	1991	1
Arkansas	Swanson v. State	823 S.W.2d	1992	1
Arkansas	Moore v. State	915 S.W.2d	1996	1
Arkansas	Whitfield v. State	56 S.W.2d	2001	1
California	People v. Soto	981 P.2d 958	1991	1
California	People v. Venegas	954 P.2d 525	1998	0
California	People v. Jones	64 P.3d 762	2003	1
Colorado	Fishback v. People	851 P.2d 884	1993	1
Colorado	Lindsey v. People	892 P.2d 281	1995	1
Colorado	People v. Shrek	22 P.2d 68	2001	1
Connecticut	State v. Hammond	604 A.2d 793	1992	0
Connecticut	State v. Skipper	637 A.2d 1101	1994	0

(*continued*)

State	Case	Citation	Year	Outcome
Connecticut	State v. Siviri	646 A.2d 169	1994	1
Connecticut	State v. Pappas	776 A.2d 1091	2001	1
Delaware	Nelson v. State	628 A.2d 69	1993	0
Delaware	Howard v. State	704 A.2d 278	1998	1
Florida	Robinson v. State	610 So.2d 1288	1992	1
Florida	Hayes v. State	660 So.2d 257	1995	0
Florida	State v. Vargas	667 So.2d 175	1995	1
Florida	Henyard v. State	689 So.2d 239	1996	1
Florida	Brim v. State	695 So.2d 268	1997	1
Florida	Murray v. State	692 So.2d 157	1997	0
Florida	Murray v. State	838 So.2d 1073	2002	0
Georgia	Caldwell v. State	393 S.E.2d 436	1990	1
Georgia	Johnson v. State	461 S.E.2d 209	1995	1
Georgia	Monroe v. State	528 S.E.2d 504	2000	1
Hawaii	State v. Montalbo	828 P.2d 1274	1992	1
Idaho	State v. Horsley	792 P.2d 945	1990	0
Idaho	State v. Faught	908 P.2d 566	1995	1
Illinois	People v. Moore	662 N.E.2d 1215	1996	1
Illinois	People v. Miller	670 N.E.2d 721	1996	1
Illinois	People v. Hickey	687 N.E.2d 910	1997	1
Indiana	Hopkins v. State	579 N.E.2d 1297	1991	1
Indiana	Davidson v. State	580 N.E.2d 238	1991	1
Indiana	Woodcox v. State	591 N.E.2d 1019	1992	1
Indiana	Jenkins v. State	627 N.E.2d 789	1993	1
Indiana	Jervis v. State	679 N.E.2d 875	1997	1
Indiana	Ingram v. State	699 N.E.2d 261	1998	1
Indiana	Smith v. State	702 N.E.2d 668	1998	1
Iowa	State v. Brown	470 N.W.2d 30	1991	1
Iowa	State v. Williams	574 N.W.2d 293	1998	1

(*continued*)

STATE SUPREME COURT CASES FOR FORENSIC DNA 167

State	Case	Citation	Year	Outcome
Kansas	Smith v. Deppish	807 P.2d 144	1991	1
Kansas	State v. Dykes	847 P.2d 1214	1993	1
Kansas	State v. Hill	895 P.2d 1238	1995	1
Kansas	State v. Colbert	896 P.2d 1089	1995	1
Kansas	State v. Isley	936 P.2d 275	1997	1
Kansas	State v. Valdez	977 P.2d 242	1999	1
Kentucky	Harris v. Commwth	846 S.W.2d 678	1992	1
Kentucky	Mitchell v. Commwth	908 S.W.2d 100	1995	1
Kentucky	Commwth v. Petrey	945 S.W.2d 417	1997	1
Kentucky	Sholler v. Commwth	969 S.W.2d 706	1998	1
Kentucky	Fugate v. Commwth	993 S.W.2d 931	1999	1
Louisiana	State v. Quatrevingt	670 So.2d 197	1996	0
Louisiana	State v. Edwards	750 So.2d	1999	1
Louisiana	State v. Hoffman	768 So.2d 542	2000	1
Maine	State v. Fleming	698 A.2d 503	1997	1
Maryland	Armstead v. State	673 A.2d 221	1996	1
Maryland	Williams v. State	679 A.2d 1106	1996	0
Maryland	Gross v. State	809 A.2d 627	2002	1
Massachusetts	Commwth. v. Curin	565 N.E.2d 440	1991	0
Massachusetts	Commwth. v. Lanigan	596 N.E.2d 272	1992	0
Massachusetts	Commwth. v. Daggett	622 N.E.2d 272	1993	0
Massachusetts	Commwth. v. Lanigan	641 N.E.2d 1342	1994	1
Massachusetts	Commwth. v. Rosier	685 N.E.2d 739	1997	1
Massachusetts	Commwth. v. Vao Sok	683 N.E.2d 671	1997	1
Massachusetts	Commwth. v. Fowler	685 N.E.2d 746	1997	1
Massachusetts	Commwth. v. McNickles	753 N.E.2d 131	2001	1
Minnesota	State v. Schwartz	447 N.W.2d 422	1989	0
Minnesota	State v. Nielsen	467 N.W.2d 615	1991	0
Minnesota	State v. Jobe	486 N.W.2d 407	1992	1
Minnesota	State v. Johnson	498 N.W.2d 10	1993	1

(continued)

State	Case	Citation	Year	Outcome
Minnesota	State v. Bloom	516 N.W.2d 159	1994	1
Minnesota	State v. Roman Nose	649 N.W.2d 815	2002	0
Minnesota	State v. Traylor	656 N.W.2d 885	2003	1
Minnesota	State v. Miller	666 N.W.2d 703	2003	1
Mississippi	Jenkins v. State	607 So.2d 1171	1992	0
Mississippi	Polk v. State	612 So.2d 381	1992	1
Mississippi	Hull v. State	687 So.2d 708	1996	1
Mississippi	Crawford v. State	716 So.2d 1028	1998	1
Mississippi	Watts v. State	733 So.2d 214	1999	1
Mississippi	Hughes v. State	735 So.2d 238	1999	1
Mississippi	Baldwin v. State	757 So.2d 227	2000	1
Missouri	State v. Davis	814 S.W.2d 543	1991	1
Missouri	State v. Kinder	942 S.W.2d 313	1996	1
Montana	State v. Moore	885 P.2d 457	1994	1
Montana	State v. Weeks	891 P.2d 477	1995	1
Montana	State v. Ayers	68 P.3d 768	2003	1
Nebraska	State v. Houser	490 N.W.2d 168	1992	0
Nebraska	State v. Carter	524 N.W.2d 763	1994	0
Nebraska	State v. Freeman	571 N.W.2d 276	1997	1
Nebraska	State v. Jackson	582 N.W.2d 317	1998	1
Nevada	Brown v. State	934 P.2d 235	1997	1
Nevada	Bolin v. State	960 P.2d 784	1998	1
New Hamp.	State v. Vandebogart	616 A.2d 483	1992	0
New Hamp.	State v. Vandebogart	652 A.2d 671	1994	1
New Hamp.	State v. Thompson	825 A.2d 490	2003	1
New Jersey	State v. Harvey	699 A.2d 596	1997	1
New Mexico	State v. Duran	881 P.2d 48	1994	1
New Mexico	State v. Anderson	881 P.2d 29	1994	1
New Mexico	State v. Stills	957 P.2d 51	1998	1

(*continued*)

State	Case	Citation	Year	Outcome
New York	People v. Wesley	633 N.E.2d 451	1994	1
N. Carolina	State v. Pennington	393 S.E.2d 847	1990	1
North Dakota	State v. Burke	606 N.W.2d 108	2000	1
Ohio	State v. Pierce	597 N.E.2d 107	1992	1
Oklahoma	Taylor v. State	889 P.2d 319	1995	1
Oklahoma	Wood v. State	959 P.2d 1	1998	1
Oklahoma	Young v. State	992 P.2d 332	1998	1
Oregon	State v. Lyons	924 P.2d 802	1996	1
Pennsylvania	Commwth. v. Crews	640 A.2d 395	1994	0
Pennsylvania	Commwth. v. Blasioli	713 A.2d 1117	1998	1
Rhode Island	State v. Morel	676 A.2d 1374	1996	1
S. Carolina	State v. Ford	392 S.E.2d 781	1990	1
S. Carolina	State v. Dinkins	462 S.E.2d 59	1995	1
S. Carolina	State v. Register	476 S.E.2d 153	1996	1
S. Carolina	State v. Council	515 S.E.2d 508	1999	1
S. Carolina	State v. Ramsey	550 S.E.2d 294	2001	1
South Dakota	State v. Wimberly	467 N.W.2d 499	1991	1
South Dakota	State v. Schweitzer	533 N.W.2d 156	1995	1
South Dakota	State v. Moeller	548 N.W.2d 465	1996	1
South Dakota	State v. Loftus	573 N.W.2d 167	1997	1
Tennessee	State v. Begley	956 S.W.2d 471	1997	1
Tennessee	State v. Scott	33 S.W.3d 746	2000	1
Texas	Kelly v. State	824 S.W.2d 568	1992	1
Texas	Fuller v. State	827 S.W.2d 919	1992	1
Texas	Campbell v. State	910 S.W.2d 475	1995	1
Texas	Massey v. State	933 S.W.2d 141	1996	1
Texas	Jackson v. State	17 S.W.3d 664	2000	1
Utah	State v. Butterfield	27 P.3d 1133	2001	1

(continued)

State	Case	Citation	Year	Outcome
Vermont	State v. Passino	640 A.2d	1994	1
Vermont	State v. Streich	658 A.2d 38	1995	0
Virginia	Spencer v. Commwth.	384 S.E.2d 785	1989	1
Virginia	Spencer v Commwth.	384 S.E.2d 775	1989	1
Virginia	Spencer v. Commwth.	393 S.E.2d 609	1990	1
Washington	State v. Cathron	846 P.2d 502	1993	0
Washington	State v. Russell	882 P.2d 747	1994	1
Washington	State v. Gentry	888 P.2d 1105	1995	1
Washington	State v. Copeland	922 P.2d 1304	1996	1
Washington	State v. Cannon	922 P.2d 1293	1996	1
Washington	State v. Jones	922 P.2d 806	1996	1
Washington	State v. Buckner	941 p.2d 667	1997	1
Washington	State v. Gore	21 P.3d 262	2001	1
West Virginia	State v. Woodhall	385 S.W.2d	1989	1
Wisconsin	State v. Hicks	549 N.W.2d 435	1996	1
Wyoming	Rivera v. State	840 P.2d 933	1992	1
Wyoming	Springfield v. State	860 P.2d 435	1993	1

Key: 0 = Inadmissible; 1 = Admissible

APPENDIX B

STATE SUPREME COURT DECISIONS FOR POLYGRAPH EVIDENCE

State	Case	Citation	Year	Outcome
Alabama	Ex Parte Dolvin	391 So.2d 677	1980	2
Alabama	Ex Parte Hinton	548 So.2d 562	1989	0
Alaska	Gafford v. State	440 P.2d 405	1968	2
Alaska	Pulakis v. State	476 P.2d 474	1970	1
Arizona	State v. Valdez	371 P.2d 894	1962	3
Arizona	State v. Treadaway	568 P.2d 1061	1977	3
Arizona	State v. Rodriguez	921 P.2d 643	1996	3
Arizona	State v. Harrod	26 P.3d 492	2001	3
Arkansas	State v. Bullock	557 S.W.2d 193	1977	3
California	People v. Carter	312 P.2d 665	1957	2
California	Ballard v. Superior Ct.	410 P.2d 838	1966	2
California	People v. Harris	767 P.2d 619	1989	2
California	People v. Jackson	920 P.2d 1254	1996	2
Colorado	People v. Anderson	637 P.2d 354	1981	0
Colorado	People v. Dunlap	975 P.2d 723	1999	0
Connecticut	State v. Mitchell	362 A2d 808	1975	0
Connecticut	State v. Esposito	670 A.2d 301	1995	0
Connecticut	State v. Porter	698 A.2d 739	1997	0

(*continued*)

State	Case	Citation	Year	Outcome
Delaware	Foraker v. State	394 A.2d 208	1978	2
Delaware	Whalen v. State	434 A.2d 1346	1980	3
Florida	Kaminski v. State	63 So.2d 339	1952	2
Florida	Codie v. State	313 So.2d 754	1975	4
Florida	Farmer v. Ft. Laudl.	427 So.2d 187	1983	3
Georgia	Stack v. State	214 S.E.2d 514	1975	2
Georgia	Scott v. State	230 S.E.3d 857	1976	1
Georgia	State v. Chambers	239 S.E.2d 324	1977	4
Hawaii	State v. Chang	374 P.2d 5	1962	2
Idaho	In re X	714 P.2d 13	1986	5
Idaho	State v. Fain	774 P.2d 252	1989	3
Idaho	State v. Travis	867 P.2d 234	1994	5
Idaho	State v. Trevino	980 P.2d 552	1999	5
Illinois	People v. Zazzetta	189 N.E.2d 260	1963	2
Illinois	People v. Baynes	430 N.E.2d 1070	1981	1
Illinois	People v. Szabo	447 N.E.2d 193	1983	2
Indiana	Zupp v. State	283 N.E.2d 942	1972	2
Indiana	Evans v. State	489 N.E.2d 942	1986	3
Indiana	Sanchez v. State	675 N.E.2d 306	1996	4
Indiana	Hubbard v. State	742 N.E.2d 919	2001	3
Iowa	State v. McNamara	104 N.W.2d 568	1960	4
Iowa	State v. Connor	241 N.W.2d 447	1976	3
Kansas	State v. Lowry	185 P.2d 147	1947	2
Kansas	State v. Lassley	545 P.2d 383	1976	3
Kansas	State v. Nemechek	576 P.2d 682	1978	3
Kansas	State v. Lumley	977 P.2d 914	1999	4
Kansas	State v. Wakefield	977 P.2d 941	1999	3
Kansas	State v. Shively	999 P.2d 952	2000	3
Kentucky	Colbert v. Commwlth.	306 S.W.2d 825	1957	2
Kentucky	Conley v. Commwlth.	382 S.W.2d 865	1964	2
Kentucky	Morton v. Commwlth.	817 S.W.2d 218	1991	1

(*continued*)

State	Case	Citation	Year	Outcome
Louisiana	State v. Refuge	270 So.2d 842	1972	2
Louisiana	State v. Corbin	285 So.2d 234	1973	1
Louisiana	State v. Governor	331 So.2d 443	1976	2
Louisiana	State v. Humphrey	455 So.2d 1155	1984	2
Louisiana	State v. Cosey	779 So.2d 675	2000	2
Maine	State v. Casale	150 Me. 310	1954	2
Maine	State v. Mottram	158 Me. 35	1962	2
Maryland	Kelley v. State	418 A.2d 217	1980	2
Massachusetts	Commwlth. v. Fatalo	191 N.E.2d 479	1963	2
Massachusetts	Commwlth. v. A Juvinile	313 N.E.2d 120	1974	4
Massachusetts	Commwlth. v. Vitello	381 N.E.2d 582	1978	3
Massachusetts	Commwlth. v. Dilego	439 N.E.2d 807	1982	3
Massachusetts	Commwlth. v. Mendes	547 N.E.2d 35	1989	1
Michigan	People v. Becker	2 N.W.2d 503	1942	2
Michigan	People v. Davis	72 N.W.2d 269	1955	2
Michigan	People v. Frechette	155 N.W.2d 830	1968	2
Michigan	People v. Barbara	255 nW.2d 171	1977	2
Minnesota	State v. Kolander	52 N.W.2d 458	1952	2
Minnesota	State v. Anderson	113 N.W.2d 4	1962	2
Minnesota	State v. Anderson	379 N.W.2d 70	1985	2
Mississippi	Hawkins v. State	77 So.2d 263	1955	2
Mississippi	Mattox v. State	128 So.2d 368	1961	2
Mississippi	Carr v. State	655 So.2d 824	1995	2
Mississippi	Gleeton v. State	716 So.2d 1083	1998	2
Missouri	State v. Cole	188 S.W.2d 43	1945	2
Missouri	State v. Stidham	305 S.W.2d 7	1957	2
Missouri	State v. Fields	434 S.W.2d 507	1968	4
Missouri	State v. Biddle	599 S.W.2d 182	1980	1
Montana	State v. Hollywood	358 P.2d 437	1960	2
Montana	State v. Campbell	579 P.2d 1231	1978	2

(*continued*)

State	Case	Citation	Year	Outcome
Montana	State v. McClean	587 P.2d 20	1978	2
Montana	State v. Bashor	614 P.2d 470	1980	2
Montana	State v. McPherson	771 P.2d 120	1989	2
Montana	State v. Staat	811 P.2d 1261	1991	2
Nebraska	Boeche v. State	37 N.W.2d 593	1949	2
Nebraska	State v. Steinmark	239 N.W.2d 495	1976	2
Nebraska	State v. Allen	560 N.W.2d 829	1997	2
Nevada	Warden v. Lischko	523 P.2d 6	1974	2
Nevada	Corbett v. State	584 P.2d 704	1978	4
Nevada	Santillanes v. State	714 P.2d 184	1986	3
Nevada	Buschauer v. State	804 P.2d 1046	1990	3
Nevada	Jackson v. State	997 P.2d 121	2000	3
New Hamp.	State v. LaForest	207 A.2d 429	1965	2
New Hamp.	State v. Stewart	364 A.2d 621	1976	2
New Hamp.	State v. French	403 A.2d 424	1979	2
New Jersey	State v. Walker	181 A.2d 1	1962	2
New Jersey	State v. McDavitt	297 A.2d 849	1972	4
New Mexico	State v. Trimble	362 P.2d 788	1961	2
New Mexico	State v. Lucerno	526 P.2d 1091	1974	4
New Mexico	State v. Dorsey	539 P.2d 204	1975	5
New Mexico	Tafoya v. Baca	702 P.2d 204	1985	5
New Mexico	State v. Sanders	872 P.2d 870	1994	5
New Mexico	State v. Harrison	7 P.3d 478	2000	5
New York	People v. Leon	255 N.E.2d 696	1969	2
New York	People v. Angelo	666 N.E.2d 1333	1996	2
N. Carolina	State v. Foye	120 S.E.2d 169	1961	3
N. Carolina	State v. Brunson	215 S.E.2d 94	1975	3
N. Carolina	State v. Milano	256 S.E.2d 154	1979	4
N. Carolina	State v. Grier	300 S.E.2d 31	1983	1
N. Carolina	State v. Brewington	532 S.E.2d 496	2000	1

(*continued*)

State	Case	Citation	Year	Outcome
North Dakota	State v. Pusch	46 N.W.2d 508	1950	2
North Dakota	State v. Swanson	225 N.W.2d 283	1974	2
North Dakota	State v. Olmstead	261 N.W.2d 880	1978	2
North Dakota	State v. Newman	409 N.W.2d 79	1987	2
North Dakota	State v. Weatherspoon	583 N.W.2d 391	1998	2
Ohio	State v. Souel	372 N.E.2d 1318	1978	4
Ohio	Criss v. Springfield Twp	564 N.E.2d 440	1990	5
Oklahoma	Henderson v. State	230 P.2d 495	1951	2
Oklahoma	Castleberry v. State	522 P.2d 257	1974	3
Oklahoma	Fulton v. State	541 P.2d 871	1975	1
Oregon	In re Black	444 P.2d 929	1968	2
Oregon	State v. Green	531 P.2d 245	1975	2
Oregon	State v. Brown	687 P.2d 751	1984	2
Oregon	State v. Lyon	744 P.2d 231	1987	1
Pennsylvania	Commwlth. v. Saunders	125 A.2d 442	1956	2
Pennsylvania	Commwlth. v. Brooks	309 A.2d 732	1973	2
Pennsylvania	Commwlth. v. Gee	354 A.2d 875	1976	2
Pennsylvania	Commwlth. v. Brockington	455 A.2d 627	1983	1
Rhode Island	Powers v. Carvalho	281 A.2d 298	1971	2
Rhode Island	State v. Dery	545 A.2d 1014	1988	2
Rhode Island	In re Odell	672 A.2d 457	1996	2
S. Carolina	State v. Britt	111 S.E.2d 669	1959	2
S. Carolina	State v. Haulcomb	195 S.E.2d 601	1973	2
S. Carolina	State v. Council	515 S.E.2d 508	1999	2
South Dakota	State v. O'Connor	194 N.W.2d 246	1972	2
South Dakota	State v. Watson	248 N.W.2d 398	1976	4
South Dakota	State v. Muetze	368 N.W.2d 575	1985	2
South Dakota	State v. Waff	373 N.W.2d 18	1985	2
South Dakota	Satter v. Solem	458 N.W.2d 762	1990	4

(*continued*)

State	Case	Citation	Year	Outcome
Tennessee	Marable v. State	313 S.W.2d 451	1958	2
Tennessee	State v. Hartman	42 S.W.3d 44	2001	2
Texas	Peterson v. State	247 S.W.2d 110	1951	2
Texas	Romero v. State	493 S.W.2d 206	1973	1
Texas	Fernandez v. State	564 S.W.2d 771	1978	2
Utah	State v. Jenkins	523 P.2d 1232	1974	4
Utah	State v. Abel	600 P.2d 994	1979	3
Utah	State v. Collins	612 P.2d 775	1980	3
Utah	State v. Rebeterano	681 P.2d 1265	1984	4
Utah	State v. Tillman	750 P.2d 54	1987	3
Utah	State v. Crosby	927 P.2d 638	1996	3
Utah	State v. Brown	948 P.2d 337	1997	3
Vermont	State v. Hamlin	499 A.2d 45	1985	2
Virginia	Lee v. Commwlth.	105 S.E.2d 152	1958	2
Virginal	Robinson v. Commwlth.	341 S.E.2d 159	1986	2
Washington	State v. Rowe	468 P.2d 1000	1970	2
Washington	State v Woo	527 P.2d 271	1974	2
Washington	State v. Renfro	639 P.2d 737	1982	4
Washington	State v. Grisby	647 P.2d 6	1982	3
West Virginia	State v. Frazier	252 S.E.2d 39	1979	1
West Virginia	State v. Beard	461 S.E.2d 486	1995	1
Wisconsin	State v. Bohner	246 N.W.314	1933	2
Wisconsin	State v. Stanislawski	216 N.W.2d 8	1974	4
Wisconsin	McLemore v. State	275 N.W.2d 692	1979	4
Wisconsin	State v. Dean	307 NW.2d 628	1981	1
Wyoming	Cullin v. State	565 P.2d 445	1977	4

Key: 0 = Inadmissible per se; 1 = Inadmissible even with a stipulation; 2 = Inadmissible; 3 = Admissible but need a stipulation; 4 = Admissible with stipulation; 5 = Admissible as a matter of judicial discretion

APPENDIX C

STATE SUPREME COURT DECISIONS FOR SYNDROME EVIDENCE

State	Case	Citation	Year	Outcome
Rape Trauma Syndrome				
Arizona	State v. Huey	699 P.2d 1290	1985	1
California	People v. Bledsoe	681 P.2d 291	1984	0
Colorado	People v. Hampton	946 P.2d 947	1987	1
Connecticut	State v. Ali	660 P.2d 337	1995	1
Illinois	People v. Wheeler	602 N.W.2d	1992	1
Indiana	Simmons v. State	504 N.E.2d 575	1987	1
Iowa	State v. Gettier	438 N.W.2d 1	1989	1
Kansas	State v. Villanueva	49 P.3d 481	2002	0
Kansas	State v. Willis	888 P.2d 839	1995	1
Kansas	State v. McQuillen	721 P.2d 740	1986	1
Kansas	State v. McQuillen	689 P.2d 822	1984	1
Kansas	State v. Bressman	689 P.2d 901	1984	0
Kansas	State v. Mark	647 P.2d 1292	1982	1
Maryland	Hutton v. State	663 A.2d 1289	1995	0
Maryland	State v. Allewalt	577 A.2d 741	1986	1
Massachusetts	Commwlth. v. Mamay	553 N.E.2d 945	1990	1
Minnesota	State v. Saldana	324 N.W.2d 227	1982	0
Minnesota	State v. McGee	324 NW.2d 232	1982	0
Missouri	State v. Taylor	663 S.W.2d	1984	0

(*continued*)

State	Case	Citation	Year	Outcome
Montana	State v. Brodinak	718 P.2d 322	1986	0
Montana	State v. Liddell	685 P.2d 618	1984	1
New Mexico	State v. Alberico	861 P.2d 192	1993	0
New York	People v. Bennett	593 N.W.2d 279	1992	1
New York	People v. Taylor	552 N.E.2d 131	1990	0
N. Carolina	State v. Hall	412 S.E.2d 883	1992	0
Pennsylvania	Commwlth. v. Gallagher	547 A.2d 335	1988	0
S. Carolina	State v. Schumbert	435 S.E.2d 859	1993	1
South Dakota	State v. Svihl	490 N.W.2d 269	1992	1
South Dakota	State v. Bachman	446 N.W.2d 271	1989	1
Vermont	State v. Kinney	762 A.2d 833	2000	1
Washington	State v. Black	745 P.2d 12	1987	0

State	Case	Citation	Year	Outcome
Battered Woman Syndrome				
California	People v. Coffman	96 P.3d 30	2004	1
California	People v. Brown	94 P.3d 574	2004	1
Connecticut	State v. Borrelli	629 A.2d 1105	1993	1
Georgia	Chapman v. State	361 S.E.2d 541	1998	1
Georgia	Smith v. State	277 S.E.2d 678	1981	1
Kansas	State v. Stewart	763 P.2d 572	1988	0
Kansas	State v. Hodges	734 P.2d 1161	1987	1
Kansas	State v. Hodges	716 P.2d 563	1986	1
Kentucky	Springer v. Commwlth.	998 S.W.2d 439	1999	1
Kentucky	Commwlth. v. Craig	783 S.W.2d 588	1990	1
Massachusetts	Commwlth. v. Conaghan	740 N.E.2d 956	2000	1
Massachusetts	Commwlth. v. Pike	726 N.E.2d 940	2000	1
Michigan	People v. Christel	537 N.W.2d 194	1995	1
Minnesota	State v. Grescinger	569 N.W.2d 189	1997	1

(*continued*)

State	Case	Citation	Year	Outcome
Minnesota	State v. Hennum	441 N.W.2d 793	1989	0
Mississippi	Lentz v. State	604 So.2d 243	1992	0
Montana	State v. Dannels	734 P.2d 188	1987	0
Nevada	Boykins v. State	995 P.2d 474	2000	0
New Jersey	State v. Kelly	478 A.2d 364	1984	1
North Dakota	State v. Liedholm	334 N.W.2d 811	1983	1
Ohio	State v. Koss	551 N.E.2d 970	1990	1
Ohio	State v. Thomas	423 N.E.2d 137	1981	0
Oklahoma	Bechtel v. State	840 P.2d 1	1992	1
Pennsylvania	Commwlth v. Dillon	598 A.2d 963	1991	1
Rhode Island	McMaugh v. State	612 A.2d 725	1992	1
S. Carolina	Robinson v. State	417 S.E.2d 88	1992	1
S. Carolina	State v. Hill	339 S.E.2d 121	1986	1
South Dakota	State v. Burtzlaff	493 N.W.2d 1	1992	1
Texas	Fielder v. State	683 S.W.2d 565	1985	0
Vermont	Soutiere v. Soutierie	657 A.2d 206	1995	1
Washington	State v. Riker	869 P.2d 43	1994	0
Washington	State v. Ciske	751 P.2d 1165	1988	1
Washington	State v. Kelly	685 P.2d 564	1984	1
Washington	State v. Allery	628 P.2d 564	1984	1
West Virginia	State v. Wyatt	482 S.E.2d 147	1996	0
Wyoming	Duran v. State	990 P.2d 1005	1999	0
Wyoming	Ryan v. State	988 P.2d 46	1999	0
Wyoming	Trujillo v. State	953 P.2d 1182	1998	1
Wyoming	Buhrle v. State	627 P.2d 1374	1981	0

Key: 0 = Inadmissible; 1 = Admissible

NOTES

CHAPTER 2 — CLUES TO JUDICIAL BEHAVIOR

1. FRE 702 states that scientific evidence must be found to be "reliable and relevant" to be admissible.

CHAPTER 4 — LIE DETECTION

1. The common application of this strategy was the placement of a tack in the subject's shoe. During questioning the subject would push down on the tack to increase physiological response.
2. "Several companies market a computerized polygraph, and such polygraphs use an algorithm developed either by Raskin and John C. Kirchner, Ph.D., or by Johns Hopkins University. The software uses the algorithm to evaluate the physiological data gathered. Davis used a Stoelting computerized polygraph with the Raskin-Kircher algorithm. After the test is completed, the polygraph charts are evaluated to determine if the pattern of physiological reactions indicates truthfulness or deception to the relevant questions. This is done by systematically comparing the strength of reactions to the relevant and control questions to determine if the reactions were consistently stronger to one or the other. Critiques about the subjective nature of polygraph scoring often center on the possibility that individual examiners may assign different scores to the same tests. Computerized scoring aims at diminishing this subjective element. The Raskin-Kircher computerized polygraph system records the data, and then the computer program analyzes the data and provides a score. The results are expressed in terms of the probability that the subject was truthful or deceptive in answering the relevant questions" (*State v. Shivley* 999 P.2d 952 [Kan. 2000], 958).
3. The discussion in this section borrows heavily from a thorough analysis by the Connecticut Supreme Court in *State v. Porter* 698 A.2d 739 (1997).
4. Post-conviction relief has also led defendants to embrace DNA evidence. In all other aspects of criminal processing, however, especially those cases reaching state supreme courts, DNA evidence was supported by the prosecution.
5. These cases were collected using a state-by-state LexisNexis search of all high court cases containing the phrase "lie detector" or the word "polygraph." Each case was then

briefly read (using the KWIC function in Lexis) to isolate wherein the court ruled on the validity/reliability and admissibility of the science used in polygraph analysis. Cases in which a court merely cited precedent were not included in the analysis. These cases do not represent significant revisiting of the question and are usually brief dismissals of a polygraph question raised amid a range of issues on appeal. Thus, only cases where the court agreed to review the question again or for the first time are included. These cases are not a sample; they compose the complete picture of court activity.

6. By comparison, federal appellate courts grant trial judges more leeway to admit, especially in the wake of *Daubert*. The majority of federal courts of appeals do not have a per se rule that polygraph evidence is inadmissible at trial. Only the fourth and the DC court maintain a per se rule of inadmissibility.

7. "To this day, the scientific community remains extremely polarized about the reliability of polygraph techniques" (*United States v. Scheffer* [1998], 1265).

8. It is true that DNA evidence has been advocated by defendants in post-conviction relief proceedings. This occurred, however, only after the scientific validity of the evidence was established in other contexts. When it came to trial court evidence, the prosecution was the only party introducing DNA evidence.

CHAPTER 5 — SYNDROME EVIDENCE

1. Variants on BWS were also searched. These included: *battered wife syndrome, battered spouse syndrome,* and *spousal abuse syndrome.*

2. This point is discussed in Giannelli and Imwinkelried (1993, 258). Legal scholars have understood the problematic nature of treating syndrome evidence in a manner similar to lab science.

3. Multivariate analysis of syndrome evidence was attempted in several iterations with no success. The small number of cases likely prevented satisfactory generation of results.

BIBLIOGRAPHY

Advanced Science and Technology Adjudication Resource Center. 2007. *Congressionally Mandated National Resource Judge Program.* http://www.einshac.org/judgeProgram.htm.

Advancing Justice Through DNA Technology. 2003. White House Report.

The American Bench: Judges of the Nation. 1977–1994. Sacramento: Forester-Long.

Allen, David W. 1991. "Voting Blocs and the Freshman Justice on State Supreme Courts." *Western Political Quarterly* 44:727–747.

American Prosecutors Research Institute. 2003. *DNA Fundamentals for the Prosecutor—Be Not Afraid.* http://www.ndaa. org/apri/programs/dna/dna_home.html.

———. 2004. *DNA Evidence Policy Considerations for the Prosecutor.* http://www.ndaa.org/pdf/dna_evidence_policy_considerations_2004.pdf.

Atkins, Burton M., and Henry R. Glick. 1976. "Environmental and Structural Variables as Determinants of Issues in State Courts of Last Resort." *American Journal of Political Science* 20:97–115.

Bander, Yigal. 1997. "United States v. Posado: The Fifth Circuit Applies Daubert to Polygraph Evidence." *Louisiana Law Review* 57:691–714.

Baum, Lawrence. 1977. "Policy Goals in Judicial Gatekeeping: A Proximity Model of Discretionary Jurisdiction." *American Journal of Political Science* 21:13–35.

Baum, Lawrence, and Bradley Canon. 1981. "Patterns of Adoptions of Tort Law Innovations: An Application of Diffusion Theory to Judicial Doctrines." *American Political Science Review* 75:975–987.

Bennett, Colin J. 1991. "What Is Policy Convergence and What Causes It?" *British Journal of Political Science* 21:215–233.

———. 1998. "Different Processes, One Result: The Convergence of Data Protection Policy in Europe and the United States." *Governance* 1:415–441.

Brace, Paul R., and Melinda Gann Hall. 1989. "Order in the Courts: A Neo-institutional Approach to Judicial Consensus." *Western Political Quarterly* 42:391–407.

———. 1990. "Neo-institutionalism and Dissent in the State Supreme Courts." *Journal of Politics* 52:54–70.

―――. "Toward an Integrated Model of Judicial Voting Behavior." *American Politics Quarterly* 20:147–168.

―――. 1993. "Integrated Models of Judicial Dissent." *Journal of Politics* 55:914–935.

―――. 1995. "Studying Courts Comparatively: The View from the American States." *Political Research Quarterly* 48:5–29.

―――. 1997. "The Interplay of Preferences, Case Facts, Context, and Rules in the Politics of Judicial Choice." *Journal of Politics* 59:1206–1231.

Brace, Paul R., Laura Langer, and Melinda Gann Hall. 2000. "Measuring the Preferences of State Supreme Court Judges." *Journal of Politics* 62:387–413.

Burgess, Ann, and Lynda Holmstrom. 1974. "Rape Trauma Syndrome." *American Journal of Psychiatry* 131:981–986.

Caldiera, Larry. 1985. "The Transmission of Legal Precedent: A Study of State Supreme Courts." *American Political Science Review* 79:178–194.

Canon, Bradley C. 1973. "Reactions of State Supreme Courts to a U.S. Supreme Court Civil Liberties Decision." *Law and Society Review* 8:109–134.

Canon, Bradley C., and Lawrence Baum. 1981. "Patterns of Adoption of Tort Law Innovations: An Application of Diffusion Theory to Judicial Doctrines." *American Political Science Review* 75: 975–987.

Canon, Bradley C., and Dean Jaros. 1970. "External Variables, Institutional Structure, and Dissent on State Supreme Courts." *Polity* 4:185–200.

Carmen, Ira H. 2004. *Politics in the Laboratory: The Constitution of Human Genomics.* Madison: University of Wisconsin Press.

Carnegie Commission on Science, Technology, and Government for a Changing World. 1993. *Science and Technology in Judicial Decision Making: Creating Opportunities and Meeting Challenges.* New York: Carnegie Commission on Science.

Carp, Robert A., and C. K. Rowland. 1996. *Politics and Judgment in Federal District Courts.* Lawrence: University Press of Kansas.

Caudill, David S., and Lewis H. LaRue. 2006. *No Magic Wand: The Idealization of Science and the Law.* Lanham, MD: Rowman & Littlefield.

Class Action Monitor. 2006. "Scientific Panels Should Determine Causation in Mass Tort Cases." *Class Action Monitor,* February 15, 24.

Cohen, Michael, James March, and Johen Olsen. 1972. "A Garbage Can Model of Organizational Choice." *Administrative Science Quarterly* 17:1–25.

Cole, Simon, and Rachel Dioso. 2005. "Law and the Lab." *Wall Street Journal,* May 13, W13.

Cole, Simon, and Rachel Dioso-Villa. 2007. "CSI and Its Effects: Media, Juries, and the Burden of Proof." *New England Law Review* 41:435–469.

Committee on Government Operations. 1976. *The Use of Polygraphs and Similar Devices by Federal Agencies.* H.R.Rep. No. 795, 94th Congress, 2d Session 46.

Cooley, Craig M. 2007. "The CSI Effect: Its Impact and Potential Concerns." *New England Law Review* 41:471–500.

Danelski, Daniel. 1966. "Values as Variables in Judicial Decision-Making: Notes Toward a Theory." *Vanderbilt Law Review* 19:721–740.

Dann, Michael B., Valerie P. Hans, and David H. Kaye. 2006 "Can Jury Trial Innovations Improve Juror Understanding of DNA Evidence?" *NIJ Journal* 255:2–6.

Davis, Bonnie J. 1994. "Admissibility of Expert Testimony After Daubert and Foret: A Wider Gate, a More Vigilant Gatekeeper." *Louisiana Law Review* 54:1307–1334.

Difonzo, J. Herbie, and Ruth C. Stern. 2007. "Devil in a White Coat: The Temptation of Forensic Evidence in the Age of CSI." *New England Law Review* 41: 503–532.

Dudley, Robert L. 1989. "State High Court Decision Making in Pornography Cases." Paper presented at the annual meeting of the American Political Science Association, Atlanta, GA.

Emmert, Craig F. 1992. "An Integrated Case-related Model of Judicial Decision Making: Explaining Supreme Court Decisions in Judicial Review Cases." *Journal of Politics* 54:543–552.

Emmert, Craig F., and C. A. Traut. 1994. "The California Supreme Court and the Death Penalty." *American Politics Quarterly* 22:41–61.

Entertainment Law Reporter. 1984. "Kansas Supreme Court Rules that Producers of the Television Show 'Lie Detector' Are Not Entitled To Enter State Penitentiary to Videotape Polygraph Examination of Inmate Seeking to Appear on the Show." *Entertainment Law Reporter* 6, no. 1: 18.

Epstein, Lee. 1993. "Interest Group Litigation During the Rehnquist Court Era." *Journal of Law and Politics* 9:639–717.

———. 1994. "Exploring the Participation of Organized Interests in State Court Litigation." *Political Research Quarterly* 47:335–351.

Epstein, Lee, and C. K. Rowland. 1991. "Debunking the Myth of Interest Group Invincibility in the Courts." *American Political Science Review* 85:205–217.

Epstein, Robert. 2002. "Fingerprints Meet Daubert: The Myth of Fingerprint 'Science' Is Revealed." *California Law Review* 75:605–657.

Faigman, David L. 1986. "The Battered Woman Syndrome and Self-Defense: A Legal and Empirical Dissent." *Virginia Law Review* 72:619–631.

Foster, Kenneth R., and Peter W. Huber. 1997. *Judging Science*. Cambridge: MIT Press.

Freely, Malcolm M. 1989. "The Significance of Prison Condition Cases: Budgets and Regions." *Law and Society Review* 23:273–282.

Galanter, Marc. 1974. "Why the 'Haves' Come Out Ahead: Speculation on the Limits of Legal Change." *Law and Society Review* 72:911–924.

Garmisa, Steven P. 1996. "Polygraph Expert Passes Daubert test, Court Rules." *Chicago Daily Law Bulletin*, March 29.

George, Tracy E., and Lee Epstein. 1992. "On the Nature of Supreme Court Decision Making." *American Political Science Review* 86:323–337.

Gerlach, Neil. 2004. *The Genetic Imaginary: DNA in the Canadian Criminal Justice System*. Toronto: University of Toronto Press.

Giannelli, Paul, and Edward J. Imwinkelried. 1993. *Scientific Evidence*. 2nd ed. Charlottesville, VA: Michie.

Glick, Henry R. 1992. "Judicial Innovation and Policy Re-Invention: State Supreme Courts and the Right to Die." *Western Political Quarterly* 45:71–92.

Glick, Henry R., and George W. Pruet Jr. 1986. "Dissent in State Supreme Courts: Patterns and Correlates of Conflict." In *Judicial Conflict and Consensus: Behavioral Studies of American Appellate Courts*, ed. Sheldon Goldman and Charles Lamb, 199–214. Lexington: University Press of Kentucky.

Glick, Henry R., and Kenneth N. Vines. 1973. *State Court Systems*. Englewood Cliffs, NJ: Prentice Hall.

Goldman, Sheldon. 1966. "Voting Behavior on the United States Courts of Appeals, 1961–1964." *American Political Science Review* 60:374–383.

———. 1975. "Voting Behavior on the United States Courts of Appeals Revisited." *American Political Science Review* 69:491–506.

Gottschall, Jon. 1983. "Carter's Judicial Appointments: the Influence of Affirmative Action and Merit Selection on Voting on the U.S. Courts of Appeals." *Judicature* 67:165–173.

Gruhl, John. 1980. "The Supreme Court's Impact on the Law of Libel: Compliance by Lower Federal Courts." *Western Political Quarterly* 33:502–519.

———. 1981. "State Supreme Courts and the U.S. Supreme Court's Post-Miranda Rulings." *Journal of Criminal Law and Criminology* 72:886–913.

Gryski, Gerard S., Eleanor C. Main, and William J. Dixon. 1986. "Models of State High Court Decision Making in Sex Discrimination Cases." *Journal of Politics* 48: 143–155.

Haas, Kenneth C. 1981. "The 'New Federalism' and Prisoner's Rights: State Supreme Courts in Comparative Perspective." *Western Political Quarterly* 34:552–571.

———. 1982. "The Comparative Study of State and Federal Judicial Behavior Revisited." *Journal of Politics* 44:721–746.

Haire, Susan Brodie, Stephanie Lindquist, and Roger Hartley. 1999. "Attorney Expertise, Litigant Success, and Judicial Decisonmaking in the U.S. Courts of Appeals." *Law and Society Review* 33:667–685.

Hall, Melinda Gann. 1987. "Constituent Influence in State Supreme Courts: Conceptual Notes and a Case Study." *Journal of Politics* 49:1117–1124.

———. 1992. "Electoral Politics and Strategic Voting in State Supreme Courts." *Journal of Politics* 54:427–446.

———. 1995. "Justices as Representatives: Elections and Judicial Politics in the American States." *American Politics Quarterly* 23:485–503.

Hall, Melinda Gann, and Paul Brace. 1989. "Order in the Courts: A Neo-institutional Approach to Judicial Consensus." *Western Political Quarterly* 42:391–407.

———. 1992. "Toward an Integrated Model of Judicial Voting Behavior." *American Politics Quarterly* 20:147–168.

———. 1994. "The Vicissitudes of Death by Decree: Forces Influencing Capital Punishment Decision Making in State Supreme Courts." *Social Science Quarterly* 75:136–151.

———. 1996. "Justices' Responses to Case Facts: An Integrated Model." *American Politics Quarterly* 24:237–261.

Hart, Angela. 1998. "New FBI Policy Revolutionizes DNA Court Testimony." *Silent Witness* 4, no. 1: http://www.ndaa.org/publications/newsletters/silent_witness_volume_4_number_1_1998.html.

Harvard Law Review. 1997. "Improving Judicial Gatekeeping: Technical Advisors and Scientific Evidence." *Harvard Law Review* 110:941–958.

———. 2000. "Navigating Uncertainty: Gatekeeping in the Absence of Hard Science." *Harvard Law Review* 113:1467–1484.

Hofferbert, Richard. 1974. *The Study of Public Policy*. Indianapolis: Bobbs-Merrill.

Hoffman, Diane E., and Karen H. Rothenberg. 2005. "When Should Judges Admit or Compel Genetic Tests?" *Science* 310:241–242.

Jaros, Dean, and Bradley C. Canon. 1971. "Dissent on State Supreme Courts: The Differential Significance of Characteristics of Judges." *Midwest Journal of Political Science* 15:322–346.

Jasanoff, Sheila. 1995. *Science at the Bar: Law, Science, and Technology in America.* Cambridge, MA: Harvard University Press.

Judicial Yellowbook: Who's Who in Federal and State Courts. 1995–2001. New York: Leadership Directories.

"Kansas Supreme Court Rules That Producers of the Television Show 'Lie Detector' Are Not Entitled to Enter State Penitentiary to Videotape Polygraph Examination of Inmate Seeking to Appear on the Show." 1984. *Entertainment Law Reporter* 6, no. 1:18.

Kilwein, John C., and Richard A. Brisbin. 1997. "Policy Convergence in State Supreme Courts." *American Journal of Political Science* 41:122–148.

King, Kamela J. 1990–1992. *BNA Directory of State and Federal Courts, Judges, and Clerks.* Washington DC: Bureau of National Affairs.

Kingdon, John. 1984. *Agendas, Alternatives, and Public Policies.* Boston: Little, Brown.

Kiser, Larry, and Elinor Ostrom. 1982. "The Three Worlds of Action." In *Strategies of Political Inquiry*, ed. E. Ostrom, 179–222. Beverly Hills: Sage.

La Morte, Tara Marie. 2003. "Sleeping Gatekeepers: United States v. Llera Plaza and the Unreliability of Forensic Fingerprinting Evidence Under Daubert." *Albany Law Journal of Science and Technology* 14:171–214.

Murphy, Walter F. 1964. *Elements of Judicial Strategy.* Chicago: University of Chicago Press.

Nagel, Stuart. 1961. "Political Party Affiliation and Judges' Decisions." *Political Science Review* 55:843–850.

National Conference of State Legislatures. 1999. *Statutes Regarding Admissibility of DNA Evidence* 1999. http://www.ncsl.org/programs/health/genetics/DNAadmiss.htm.

National Institute of Justice. 1999. *What Every Law Enforcement Officer Should Know About DNA Evidence.* http://www.ojp.usdoj.gov/nij/pubs-sum/000614.htm.

National Research Council. 1992. *DNA Technology in Forensic Science.* Washington DC: National Academy Press.

———. 1996. *DNA Technology in Forensic Science.* Washington DC: National Academy Press.

Neufeld, Peter. 1990. "Have You No Sense of Decency?" *Journal of Criminal Law and Criminology* 84:189–202.

O'Connor, Karen. 1980. *Women's Organizations' Use of Courts.* Lexington, MA: Lexington Books.

Orne, Martin. 1975. "Implications of Laboratory Research for the Detection of Deception." In *Legal Admissibility of the Polygraph*, ed. N. Ansley, 94–119. Springfield, IL: Charles C. Thomas.

Pinsky, Lawrence S. 1997. "The Use of Scientific Peer Review and Colloquia to Assist Judges in the Admissibility Gatekeeping Mandated by Daubert." *Houston Law Review* 34:527–578.

Prichett, C. Herman. 1948. *The Roosevelt Court.* New York: Macmillan.

Raloff, Janet. 2005. "Benched Science." *Science News*, October 8, 232–234.

Raskin, David C. 1986. "The Polygraph in 1986: Scientific, Professional, and Legal Issues Surrounding Application and Acceptance of Polygraph Evidence." *Utah Law Review* 13:29–74.

Reid, Tracy V. 1988. "Judicial Policy-making and Implementation: An Empirical Examination." *Western Political Quarterly* 41:509–527.

Rhode, David W. 1972. "Policy Goals, Strategic, Choice and Majority Opinion Assignments in the U. S. Supreme Court." *Midwest Journal of Political Science* 16:652–682.

Romans, Neil T. 1974. "The Role of State Supreme Courts in Judicial Policy Making: Escobedo, Miranda, and the Use of Judicial Impact Analysis." *Western Political Quarterly* 27:38–59.

Sabatier, Paul A. 1988. "An Advocacy Coalition Framework of Policy Change and the Role of Policy-oriented Learning Therein." *Policy Sciences* 21:129–168.

———. 1991. "Toward Better Theories of the Policy Process." *PS: Political Science and Politics* 24:147–156.

Schmitt, Glenn R. 2007. "Online DNA Training Targets Lawyers, Judges." *NIJ Journal* 256: http://www.ojp.usdoj.gov/nij/journals/256/online-dna-training.html.

Schubert, Glendon. 1960. *Constitutional Politics: The Political Behavior of Supreme Court Justices and the Constitutional Policies They Make*. New York: Holt, Rinehart, and Winston.

———. 1965. *The Judicial Mind: The Attitudes and Ideologies of Supreme Court Justices, 1946–1963*. Evanston, IL: Northwestern University Press.

Schutz, Amy T. 1994. "The New Gatekeepers: Judging Scientific Evidence in a Post-Frye World." *North Carolina Law Review* 72:1060–1084.

Segal, Jeffery A. 1984. "Predicting Supreme Court Decisions Probabilistically: The Search and Seizure Cases (1962–1981)." *American Political Science Review* 78:891–900.

———. 1986. "Supreme Court Justices as Human Decision Makers: An Individual Level Analysis of Search and Seizure Cases." *Journal of Politics* 48:938–955.

———. 1988. "Amicus Briefs by the Solicitor General During the Warren and Burger Courts." *Western Political Quarterly* 41:134–144.

Segal, Jeffery A., and Albert D. Cover. 1989. "Ideological Values and Votes of U.S. Supreme Court Justices." *American Political Science Review* 83:557–565.

Segal, Jeffery A., and Harold J. Spaeth. 1993. *The Supreme Court and the Attitudinal Model*. New York: Cambridge University Press.

Shapiro, E. Donald, and Stewart Reifler. 1996. "Forensic DNA Evidence and the United States Government." *Medicine Science and the Law* 36, no. 1: 43–51.

Shapiro, Martin. 1964. *Law and Politics in the Supreme Court: New Approaches to Political Jurisprudence*. London: Free Press.

———. 1970. "Toward a Theory of Stare Decisis." *Journal of Legal Studies* 1:125–134.

Sheehan, Reginald S., William Mishler, and Donald R. Songer. 1992. "Ideology, Status, and the Differential Success of Direct Parties before the Supreme Court." *American Political Science Review* 86:464–471.

Sickels, Robert. 1965. "The Illusion of Judicial Consensus." *American Political Science Review* 59:100–104.

Songer, Donald R. 1982. "Consensual and Nonconsensual Decisions in Unanimous Opinions of the United States Courts of Appeals." *American Journal of Political Science* 26:225–239.

Songer, Donald R., and Sue Davis. 1990. "The Impact of Party and Region on Voting Decisions in the United States Court of Appeals." *Western Political Quarterly* 43:317–334.

Songer, Donald R., and Susan Haire. 1992. "Integrating Alternative Approaches to the Study of Judicial Voting: Obscenity Cases in the U.S. Courts of Appeals." *American Journal of Political Science* 36:963–982.

Songer, Donald R., and Ashlyn Kuersten. 1995. "The Success of Amici in State Supreme Courts." *Political Research Quarterly* 48:31–42.

Songer, Donald R., and Reginald S. Sheehan. 1992. "Who Wins on Appeal? Upperdogs and Underdogs in the United States Courts of Appeals." *American Journal of Political Science* 36:235–258.

———. 1993. "Interest Group Success in the Courts: Amicus Participation in the Supreme Court." *Political Research Quarterly* 46:339–354.

Songer, Donald R., Reginald S. Sheehan, and Susan B. Haire. 1999. "Do the 'Haves' Come Out Ahead Over Time? Applying Galanter's Framework to Decisions of the U.S. Courts of Appeals, 1925–1988." *Law and Society Review* 33:811–832.

Songer, Donald R., and Susan J. Tabrizi. 1999. "The Religious Right in Court: The Decision Making of Christian Evangelicals in State Supreme Courts." *Journal of Politics* 61:507–526.

Tarr, G. Alan. 1994. "The Past and Future of the New Judicial Federalism." *Publius* 24:63–79.

Tate, C. Neal. 1981. "Personal Attribute Models of the Voting Behavior of U.S. Supreme Court Justices: Liberalism in Civil Liberties and Economic Decisions, 1946–1978." *American Political Science Review* 75:355–367.

Tate, C. Neal, and Roger Handberg. 1991. "Time Binding and Theory Building in Personal Attribute Models of Supreme Court Voting Behavior, 1916–88." *American Journal of Political Science* 35:460–480.

Thompson, William C. 1997. "A Sociological Perspective on the Science of Forensic DNA Testing." *UC Davis Law Review* 30:1113–1132.

U.S. Congress, Office of Technology Assessment. 1983. *Scientific Validity of Polygraph Testing: A Review and Evaluation—A Technical Memorandum*. OTA-TM-H-15.

U.S. Department of Defense. 1984. *The Accuracy and Utility of Polygraph Testing* Washington DC: Government Printing Office.

Vose, Clement E. 1959. *Caucasians Only*. Berkeley and Los Angeles: University of California Press.

Walker, Lenore. 1979. *The Battered Woman*. New York: Harper & Row.

———. 1984. *The Battered Woman Syndrome*. New York: Harper & Row.

Warner, William J. 2005. "Polygraph Testing: A Utilitarian Tool." *FBI Law Enforcement Bulletin* 74 (April): 10–13.

Wenner, Lettie McSpadden, and Lee E. Dutter. 1989. "Contextual Influences on Court Outcomes." *Western Political Quarterly* 41:113–134.

Wheeler, Stanton, Bliss Cartwright, Robert A. Kagan, and Lawrence M. Friedman. 1987. "Do the 'Haves' " Come Out Ahead? Winning and Losing in the State Supreme Courts, 1870–1970." *Law and Society Review* 21:403–445.

INDEX

acquittal, 74
admissibility, 1, 139–140, 162–163; amici briefs, 58–60, 97–98, 130–131; appellant status, 55, 126–128; attitudinal factors, 49–54, 64–66, 91–94, 104, 122–126; attorney expertise, 55–56; battered woman syndrome, 105–133; chronology, 40–41, 64, 66, 79, 86, 89, 97, 109, 133, 149, 153; DNA, timing of, 41, 43–45; executive branch policy influence, 151–153; expert testimony, 56–58, 97, 128–129; federalism, 155–156; forensic DNA, 75; institutional and organizational support, 54–64, 65–67, 95–99, 126–132, 136–137, 161; judicial education influence, 153–155; legal factors, 45–49, 64–66, 119–122, 135; legal standards, 1, 5, 14, 18–19; media influence, 158–161; middle ground approach, *see* middle ground approach; multivariate analysis, 64–66, 99–101; partisanship, 49–54, 92–94, 103, 122–125, 135–136; patterns of, 156–158; policy advocates, 58–60, 97–98, 130–131; political factors, 54–64, 126–132; polygraphs, 68, 71, 73–82; popular culture influence, 159–161; precedent, 88; rape trauma syndrome, 105–133; regionalism, 52–54, 66, 94–95, 125–126, 135–136, 145; the role of legal facts, 121; syndrome evidence, 105–133; third-party reports, 98–99, 131–132; threshold of, 31–34. *See also* legislative admissibility
Advanced Science and Technology Adjudication Resource Center (ASTAR), 151, 154. *See also* Einstein Institute for Science, Health, and the Courts (EINSHAC)
Advancing Justice Through DNA Technology, 39, 152–153
advocates, *see* policy advocates
Alabama Supreme Court, 48, 50
Alaska Supreme Court, 148
Allen, David W., 12
American Civil Liberties Union, 59–60, 97–98
American Judges Association, 154
American Polygraphers Association (APA), 97
American Prosecutors Research Institute, 152
American Psychological Association, *Diagnostic and Statistical Manual of Mental Disorders* (3rd ed.) (DSMIII), 131–132
amicus curiae, 6, 12, 54, 58–60, 97–98, 130–131, 133, 137, 140, 149
anxiety, 69, 106
appellant status, 25–26, 35, 54–55, 126–128, 138
appellate courts, 8
appellate jurisdiction, 144
Appellate Defense Office (Michigan), 97
Arizona Supreme Court, 1–2, 4, 64, 114
Arizona v. Bible, 858 P.2d 1152, 2
Arkansas Supreme Court, 45, 82
Association of Criminal Defense Lawyers (Michigan), 97
Atkins, Burton M., 11
attitudinal model, 20–23, 31–35, 52–53, 64, 108, 132, 138–139
attorney general, 58, 144, 147. *See also* prosecution

191

Attorney General for Division of Criminal Justice (New York), 59
Attorney General of Colorado, 59
Attorney General of Minnesota, 59
Attorney General of Ohio, 59
Attorney General of Washington, 59
attorneys, 35, 54–55; expertise, 55–56. *See also* defense; prosecution

Ballard v. Superior Court, 410 P.2d 838 (Cal. 1966), 99
ballistics, 13
Bander, Yigal, 19
banding patterns, 38, 74
band-shifting analysis, 43
battered woman syndrome, 9, 23, 105–133, 157; admissibility, 5, 109–113; definition of, 107–108, expert testimony, 128–129; recantations, 107; as therapeutic tool, 108. *See also* syndrome evidence
battle of the experts, 97, 129
Baum, Lawrence, 11, 20, 26, 156
Bennett, Colin J., 26
bite-mark analysis, 13
black robes, 1–2, 4, 9–11, 43, 82, 114, 140–142
blood typing, 9
bounded rationality, 15
Brace, Paul R., 11, 17–18, 20–21, 27
Brisbin, Richard A., 26–27, 29, 155
Burgess, Ann, 105–106
Bush, George W., 39, 153. *See also* President of the United States

Caldiera, Larry, 11, 22, 26–27, 155
Caldwell v. State, 393 S.E.2d 436, 441 (1990), 46
California Public Defenders Association, 59
California Public Defenders Office, 130
California Supreme Court, 63, 114, 130
California Women Lawyers, 130–131
Canon, Bradley C., 12, 20, 26–27, 86, 115, 156
capital cases, 11, 25, 55, 75. *See also* death penalty
Carmen, Ira H., 15, 141
Carnegie Commission on Science, Technology, and Government for a Changing World, 141
Carp, Robert A., 15, 21–22, 24
Caudill, David S., 155, 161
ceiling principle, 61
Central Intelligence Agency (CIA), 70
child sexual abuse accommodation syndrome, 138, 156

clearance, 31–34
Cohen, Michael, 146
cognition, 40, 91, 118
cognitive decision making, 126, 128
cognitive framing and evaluation, 16, 30, 66, 73, 102, 108, 119, 136, 139, 158
cognitive processing, 138–139
cognitive psychology, 15–16, 99; persuasion, 56–58
cognitive satisfaction, 102, 122
Cole, Simon, 159–160
Colorado Supreme Court, 45, 84, 114
common law, 8, 18
Commonwealth v. Dillon, 598 A.2d 963, 971 (Penn. 1991), 113
Connecticut Supreme Court, 43, 52, 84–85, 114
consensual intercourse, 106, 109, 112, 121
conservative values, 108, 124
constituents, 21
Cooley, Craig M., 161
counsel, *see* attorneys; defense; prosecution
counsel expertise, 29, 32, 54–56, 74, 128
court analysis, 86–88, 100–101
court behavior, *see* judicial behavior
courts, 97; attitudinal factors, 20–22; exemplar courts, 26–27; federalism, 155–156; organizational interests, 27; partisanship, 21–22; peer courts, 26–27, 161; policy makers, 126; political preferences, 5; polygraphs, 76; regional differences, 22; syndrome evidence, 109. *See also* judges
Cover, Albert D., 20–21
credibility, 107
crime scene, 37–38, 61, 68, 71, 73
Crime Scene Investigation (*CSI*) (television program), 144; *CSI* Effect, 160–161
criminal courts, 61, 120, 146
criminal investigation, 70, 73
criminal justice, 3, 118, 122, 131, 155
Criminal Law Clinic (New York City), 59
criminal proceedings, 75–76, 102, 105–106
CSI, *see Crime Scene Investigation* (*CSI*)
cues, 15–16, 29, 95, 99, 126, 128
cycle of abuse, 107

Danelski, Daniel, 20
Dann, Michael B., 152
Daubert v. Merrill Dow Pharmaceuticals, 113 S.Ct.2786 (1993), 2, 18–19, 40, 46–49, 65, 85, 89, 100, 119–120, 146, 150
Davis, Bonnie J., 21–22
death penalty, 11, 75, 146. *See also* capital cases

INDEX

deception, 68–72
decision making, *see* human decision making
Defender's Association of Philadelphia, 130
defense, 30, 55, 67, 107, 121, 123–124, 129, 144; amici briefs, 58–60; defendants, 2, 5, 9, 14, 22, 49–50, 52, 54–55, 57, 72–75, 85, 92–96, 101–103, 108, 121–122, 126–128, 136, 138–140; defense attorneys, 25, 56, 61, 65, 98, 128, 145, 149, 153, 160
Delaware Supreme Court, 42, 53, 64
Democrats, 16, 21–22, 50–52, 54, 66, 92–94, 122–125, 132–133, 136, 138–139, 150
Department of Defense, *see* United States Department of Defense
Department of Defense Polygraph Institute, 71
Department of Justice, *see* United States Department of Justice
Difonzo, J. Herbie, 160–161
diffusion, 12, 26–27, 73, 156
Dioso-Villa, Rachel, 159–160
disorders, 107
district attorneys, 58. *See also* prosecution
District Attorneys Association of New York, 59
divergence, 88, 157
Dixon, William J., 12
DNA, *see* forensic DNA
domestic violence, 113
DSMIII, *see* American Psychological Association, *Diagnostic and Statistical Manual of Mental Disorders* (3rd ed.)
Dudley, Robert L., 21
due process, 75. *See also* fundamental fairness
Dutter, Lee E., 22
dynamic jurisdictions, 44–45, 48, 64, 82, 85, 114–115

ear-print analysis, 160
Einstein Institute for Science, Health, and the Courts (EINSHAC), 141, 151. *See also* Advanced Science and Technology Adjudication Center (ASTAR)
electrophoresis process, 38
Emmert, Craig F., 18
emotional state, 69, 72, 74, 105
Epstein, Lee, 12, 17–18, 24–25, 30, 126
Epstein, Robert, 29
Equal Justice for Women in the Courts Act of 1994 (SEC 40401), 151
exculpation, 107

executive branch, 151–153. *See also* President of the United States
Ex Parte Perry, 586 So.2d 242, 244 (Ala. 1991), 42
expert testimony, 2, 32–33; battered woman syndrome, 107–108, 113; child sexual abuse accommodation syndrome, 150; forensic DNA, 36, 54–58, 65; professionalization, 27–28; polygraphs, 72, 74–75, 88, 102; post-traumatic stress syndrome, 150; rape trauma syndrome, 106–108, 150; syndrome evidence, 108, 118, 128–129
extra-trial proceedings, 75–76, 79, 84, 87
eyewitness testimony, 86, 141

facial mapping, 160
Faigman, David L., 108
fear of detection, 106
Federal Bureau of Investigation (FBI), 39, 60, 70, 147, 151–153
federal courts, 27, 155–156
federalism, 145, 155–156
Federal Rule of Evidence (FRE) 702, 18–19, 40, 46–49, 65, 68, 89, 100–101, 119–120. *See also* rules of evidence
feminism, 104, 108, 125–126. *See also* gender politics
Fielder v. State, 683 S.W.2d 565, 592 (Tex. 1985), 113
fingerprints, 3, 9, 28, 37, 141
Florida Supreme Court, 42–43, 50, 114
forensic DNA, 2, 10, 19, 23, 28, 37–39, 151–153, 157; admissibility, 5, 36, 40–45, 73, 147; amici briefs, 58–60; appellant status, 55; attitudinal factors, 49–54, 64–66; attorney expertise, 55–56; chronology, 40–41, 64, 66, 79, 86, 89, 97, 109, 133, 149, 153; databases, 147, 152, 157; evidence, 13, 25, 36–37, 40, 43, 134; executive branch policy, 3, 152–153; expert testimony, 56–58; fragments, 38; institutional and organizational factors, 54–66, 136–137; judicial partisanship, 49–52; jurisprudence, 43–45; labs, 37, 39, 59, 67; legal factors, 45–49, 64–66, 135; legislative admissibility, 148; match statistics, 38–40, 42–43, 61, 74; media coverage, 158; multivariate analysis, 64–66; National Research Council (NRC), 58, 60–66, 131–132, 137, 140, 147, 152, 162; partisanship, 135–136, 139; policy advocates, 58–60; politics, 39–41, 54–64; polygraph evidence comparison, 5, 68–69, 72–76, 79, 82, 88, 97–98,

forensic DNA (continued)
102–104; post-conviction, 30, 157;
regionalism, 52–54, 66, 135–136; syndrome evidence comparison, 9–10, 108,
114, 119–120, 126, 128–129, 131–133;
techniques, 37–38, 40, 42, 47; testing, 13;
theory of, 37, 40, 42
forensic evidence, 105, 161
forensic labs, 29, 128
forensic science, 13, 157, 160. *See also* ballistics; bite-mark analysis; blood typing; ear-print analysis; facial mapping; fingerprints; forensic DNA; hair analysis; handwriting analysis; lip reading; Luminol treatments
forensic scientists, 28, 74, 106, 153. *See also* scientific knowledge
Foster, Kenneth, 141
Freely, Malcolm M., 22
Frye v. United States, 293 F.1013 (1923), 18, 40, 46–49, 65, 70, 89, 100–101, 115, 119–120, 135, 138, 146, 150. *See also* general acceptance test
fundamental fairness, 75, 87

Galanter, Marc, 24
"garbage can model," *see* policy streams approach
Garmisa, Steven P., 156
gatekeepers, 3–4, 6–8, 15, 34, 37, 40, 49, 50, 55, 58, 67, 76, 88, 91–92, 109, 134–142, 162–163
gatekeeping, 1, 3, 138–140, 161–162; amici briefs, 58–60, 97–98, 130–131; appellant status, 55, 126–128; attitudinal factors, 20–22, 49–54, 64–66, 91–94, 104, 122–126; attorney expertise, 55–56; chronology, 40–41, 64, 66, 79, 86, 89, 97, 109, 133, 149, 153; clues to, 14–35, 101–104, 132–133; definition of, 8; DNA, 39–40; executive branch policy influence, 151–153; expert testimony, 56–58, 97, 128–129; federalism, 155–156; institutional and organizational factors, 5, 54–67, 95–99, 126–132, 136–137, 161; judicial education influence, 153–155; judicial outcomes, 30–31, 76, 95; judicial selection, 145; legal commentary, 2, 140–142, 158–159; legal factors, 17–19, 45–49, 64–66, 88–91, 119–122, 135; legislative admissibility, 148–151; media, 158–161; model of, 31–34; multivariate analysis, 64–66, 99–101; partisanship, 49–54, 92–94, 103, 122–125, 135–136; policy, 9–12, 49, 103, 107, 143–163; policy advocates, 58–60, 97–98, 131;
political factors, 22–30, 54–64, 126–132, 134; political implications, 133; as a political phenomenon, 8, 75, 108; polygraphs, 73, 76–82; popular culture influence, 159–161; precedent, 88; puzzles of, 42, 82, 105, 128, 134, 137–138, 156, 162–163; regionalism, 22, 52–54, 66, 94–95, 125–126, 135–136, 144; role of legal facts, 121–22; syndrome evidence, 109–113; theories of, 156–158; third-party reports, 98–99, 131–132
gender politics, 108, 131, 133. *See also* feminism
general acceptance test, 19, 40, 46, 61, 68, 89, 99, 115, 119, 135. *See also Frye v. United States*
George, Tracy E., 17–18, 24–25, 29, 126
Georgia Supreme Court, 45–46, 114
Gerlach, Neil, 147
Giannelli, Paul, 13, 18, 72, 105, 132
Glick, Henry R., 11, 156
Goldman, Sheldon, 20–21
Gottschall, Jon, 21
government officials, 60, 70, 73, 75, 98, 101
Gruhl, John, 27
Gryski, Gerard S., 12

Haas, Kenneth C., 156
hair analysis, 9, 13, 138, 156
Haire, Susan Brodie, 18, 21, 24
Hall, Melinda Gann, 11, 17–18, 20–21, 27
Handberg, Roger, 20–21
handwriting analysis, 9, 13
Hans, Valerie P., 152
Hart, Angela, 152
Hartley, Roger, 24
Hawaii Supreme Court, 43, 82
high courts, *see* state supreme courts
Hofferbert, Richard, 144
Hoffman, Diane E., 155
Holmstrom, Lynda, 105–106
Horizontal Gaze Nystagmus (HGN), 13, 16, 21
Huber, Peter, 141
human decision making, 15–17, 35, 73, 102–103, 108, 118, 134, 138
Human Genome Project, 151
hypnosis, 13

Idaho Supreme Court, 4, 42, 50, 64, 79, 148
identical twins, 37
identity, 37, 72
ideology, *see* conservative values; liberal values; political ideology
Illinois Supreme Court, 45, 114

INDEX 195

imminent danger or harm, 107, 113, 122
Imwinkelried, Edward J., 72, 105, 132
Indiana Supreme Court, 42, 45, 114
industry, 60, 70, 73, 75, 98, 101
innocence, 70–72
Institute of Human Genetics, 59–60
institutional rational choice, 145–146
institutional support, 138
interest groups, 12, 60, 98, 103, 137
Iowa Supreme Court, 45, 114

Jaros, Dean, 12, 86, 115
Jasanoff, Sheila, 141–142
Journal of Forensic Science, 71
judges, 1, 7, 86, 91, 99, 102, 148, 153; appointment of, 21; as cognitive decision makers, 128; dissent, 86; gatekeeping role, 7, 15–17; 105; human decision making, 131, 134, 138; idealistic or pragmatic, 155, 161; judicial interest groups, 154; National Research Council, 61; news editorials, 21; organizational interests, 27; partisanship, 14, 49–54, 92, 103, 122; party affiliation, 21–22, 122–125, 150; policy making, 105, 148–151; political attitudes, 12, 16, 31, 102, 108, 119, 139; political campaigns, 22; political ideology, 20, 139; political preferences, 4, 14, 16, 20, 49–54; and polygraphs, 74; religious beliefs, 12; and scientific experts, 2–3, 140–141, 151; syndrome evidence, 118; trial court, 46, 122
Judges Journal, 159
judicial analysis, 118–119. *See also* court analysis
judicial behavior, 139–140, 162–163; amici briefs, 58–60, 97–98, 130–131; appellant status, 55, 126–128; attitudinal factors, 20–22, 31, 49–54, 64–66, 91–94, 104, 122–126; attorney expertise, 55–56; chronology, 40–41, 64, 66, 79, 86, 89, 97, 109, 133, 149, 153; clues to, 14–35, 132–133, 143–163; destabilization, 86, 90, 99, 103; diffusion, 10–11, 26–27; dissent, 50–51, 92, 115–118, 124; emulation, 26–27; expert testimony, 56–58, 97, 128–129; federalism, 155–156; freshman effect, 12; gatekeeping, 31–34; human decision making, 15–17, 73, 103, 108; innovation, 26, 139; institutional and organizational support, 11, 14, 23–31, 54–67, 95–99, 126–132, 136–137, 161; judicial education influences, 151, 154–155; judicial selection, 145; legal factors, 45–49, 64–66, 88–91, 119–122, 135; legal frame, 17–19; multivariate analysis, 64–66, 99–101; organized interests, 27, 29; partisanship, 5, 21–22, 49–54, 92–94, 103, 122–125, 135–136; peer courts, 26–27, 155–156; policy advocates, 58–60, 97–98, 130–131; political factors, 54–64, 126–132; political implications, 133; political preferences, 20–22; polygraphs, 76–82, 86; popular culture influences, 159–161; precedent, 88; puzzles of, 96; regionalism, 5, 22, 52–54, 66, 94–95, 125–126, 135–136; religious beliefs, 12; role of legal facts, 121; syndrome evidence, 105–133; theories of, 4–5, 8, 11, 14–15, 22–34; third-party reports, 131–132; third-party support, 98–99; uncertainty, 12
judicial debate, 86–88, 115–118, 140
judicial discretion, 79, 92, 112
judicial selection, 27, 145
Judicial Yellowbook, 50, 92, 122
juries, 72–73, 75, 86, 113, 147, 160–162
justice, 3–4, 18, 60, 103, 141

Kansas County Prosecutors and District Attorneys Association, 130
Kansas Supreme Court, 1–2, 4, 45, 71, 79, 113–115
Kansas v. Marks, 647 P.2d 1292 (1982), 2
Kaye, David H., 152
Kentucky Supreme Court, 45, 114
Kilwein, John C., 26–27, 29, 155
Kingdon, John, 146
Kiser, Larry, 145
Kuersten, Ashlyn, 12

labs, *see* forensic DNA: labs; forensic labs
lab science, 72, 108, 119
LaMorte, Tara Marie, 3
Landes, Richard J., 26
Langer, Laura, 21
LaRue, Lewis H., 155, 161
law, 4, 12, 16–17, 31–33, 35, 67, 102, 106, 120, 129, 138, 141–142, 160–161
law-and-order politics, 21, 52, 66, 93, 108, 124–125, 139, 144–145, 150, 157
law enforcement, 3, 12–14, 16, 22–23, 25, 28, 39, 52, 54, 59, 67, 70, 73–75, 79, 87, 93, 96, 98, 101, 103, 128–129, 137, 139–140, 144–145, 147, 149–153, 163; expert testimony, 57; organizational advantage, 73; political advantage, 73. *See also* prosecution
legal capital, 26
legal doctrine, 18, 145, 150

legal experts, 60, 98, 140
legal facts, 120–121, 135, 138–139
legal model, 17–18, 23, 34–35, 64–65, 120, 132. *See also* judicial behavior: legal factors
"legal mutation," 86, 103, 109, 115, 118–119, 162. *See also* middle ground approach: polygraphs
legal standards, 4–5, 18–19, 31–34, 45–49, 65–67, 88–90, 100–102, 108, 118–119, 134–135, 138–139, 150, 155, 161; as a political choice, 109; political implications, 67; strict or liberal, 15, 18–19, 46–47, 49, 65–67, 89–90, 93, 100, 102–103, 119, 135. *See also* admissibility: legal factors; *Daubert v. Merrill Dow Pharmaceuticals*, 113 S.Ct.2786 (1993); Federal Rules of Evidence (FRE) 702; *Frye v. United States*, 293 F.1013 (1923)
legislation, 108, 160
legislative admissibility, 148–151. *See also* legislatures
legislative policy making, 137
legislatures, 134, 137, 140, 146, 148–151
leverage, *see* political leverage
Levi, Edward H., 17
liberal legal standards, *see* legal standards: strict or liberal
liberal values, 108, 125
lie detectors, *see* polygraphs
Lie Detector (television program), 160
Lindquist, Stephanie, 24
lip reading, 160
litigants, 23–26, 29, 37, 49–50, 61, 72, 92
Louisiana Supreme Court, 4, 43, 50, 63, 79, 84
Luminol treatments, 13

Main, Eleanor C., 12
Maine Supreme Judicial Court, 43
Mangen Research Association Incorporated, 59
Maryland Court of Appeals, 43, 45, 50, 53, 82, 114–115, 148
Massachusetts Lawyers Weekly, 158
Massachusetts Supreme Judicial Court, 42, 48, 50, 53, 114
mass political behavior, 143–144
media coverage, 158–159, 162
Michigan Association of Criminal Defense Lawyers, 98
Michigan Supreme Court, 79, 98, 114
middle ground approach, 91, 96, 99, 122; polygraphs, 86, 90–91, 120; syndrome evidence, 122. *See also* procedural innovations

Minnesota Supreme Court, 36, 40, 42–43, 50, 64, 114–115
Mississippi Supreme Court, 42, 114
Missouri Supreme Court, 114
Mishler, William, 24–25
mitigating evidence, 75
Moment of Truth (television program), 160
momentum theory, 26
monotonic jurisdictions, 45, 83–84
Montana Supreme Court, 45, 113–114
multiple causality, 106, 112
multivariate analysis, 64–66, 99–101
murder, 23, 113, 121–122
Murphy, Walter F., 17
mutation, *see* "legal mutation"

Nagel, Stuart, 21
National Association of Parents of Murdered Children, 59
National Clearinghouse for the Defense of Battered Women, 130–131
National Conference of State Legislatures, 148
National Council of Juvenile and Family Court Judges, 154
National Criminal Justice Reference Service, 152
National Institute of Justice, 152
National Judges' Science School, 151, 154
National Legal Aid and Defender's Association, 59
National Research Council, 5, 28, 38, 53–55, 60–66, 103, 137, 147, 152, 162
National Victim Assistance Academy, 152
Native Americans, 61
Nebraska Supreme Court, 43, 64
neo-institutional perspective, 27
Neufeld, Peter, 152
Nevada Supreme Court, 45–46, 114
New Hampshire Supreme Court, 48, 50, 53, 64
New Jersey Supreme Court, 43, 114
New Mexico Supreme Court, 45, 79, 114
New York Association of Criminal Defense Lawyers, 97–98
New York Court of Appeals, 43, 98, 114
North Carolina Supreme Court, 43, 114
North Dakota Supreme Court, 43, 114
Northwest Women's Law Center, 130–131

O'Connor, Karen, 24
Office of Technology Assessment, 99, 104; *Report on Polygraphs*, 70
Ohio Supreme Court, 43, 113–115
Oklahoma Court of Criminal Appeals, 45, 114

open systems theory, 143–145
Oregon Supreme Court, 43
original intent, 17
Orne, Martin, 72
Ostrom, Elinor, 145

partisanship, 5, 21–22, 49–54, 67, 92–94, 103, 122–125, 135–136
PCR analysis, 37–39, 42–43
Pennsylvania Supreme Court, 50, 64, 114, 130
People v. Anderson, 637 P.2d 354, 360 (Colo. 1981), 70, 72
People v. Angelo, 666 N.E.2d 1333 (N.Y. 1996), 97
People v. Barbara, 255 N.W.2d 171 (Mich. 1977), 97
People v. Harris, 767 P.2d 619 (Cal. 1989), 85
perpetrator, 113, 147
per se rule, 76, 79, 87, 96, 112–113, 122, 128
physiology, 68–71, 73–74, 105
Pinsky, Lawrence S., 3
plain meaning, 17
police officers, 72, 153
policy advocates, 29–30, 35, 58–60, 95, 101–102, 104, 130–131, 137, 140
policy convergence, 45, 88, 155–156, 163
policy entrepreneurs, 149, 153
policy experimentation, 90
policy innovation, 150
policy making, theories of, 6, 134, 138, 143–148
policy outcomes, 143–163
policy streams approach (garbage can model), 6, 146–147
political attitudes, 134
political ideology, 21, 23, 40, 104, 108, 125; of judges, 20
political leverage, 23–24, 27–29, 31–35, 58, 73, 95, 103, 136, 142
political party affiliation, *see* partisanship
political science, 4, 6, 14–16, 34, 36, 103
politics, 40, 54, 67–68, 73–76, 95–99, 103–104, 108–109, 126–132, 134, 140–142
polygraph examiner, 69–73, 128; "friendly polygraphers," 72, 128; subjectivity, 135
polygraph industry, 102
polygraph institutes, 73
polygraph machine, 70
polygraphs, 13, 19, 23, 25, 28, 67–104, 106, 147, 152–153; admissibility, 5, 73–74, 76–82, 156; amici briefs, 97–98; chronology, 40–41, 64, 66, 79, 86, 89, 97, 109, 133, 149, 153; computerized scoring programs, 71; confessions, 73; countermeasures, 69; evidence, 9–10, 25, 134; expert testimony, 97; forensic DNA evidence comparison, 5, 68–69, 72–76, 79, 82, 88, 97–98, 102–104; guilty knowledge test, 71; human interpretation, 69–71; institutional and organizational factors, 73, 95–99, 136–137; jurisprudence, 82–85; legal factors, 88–91, 100–101, 135; legislative admissibility, 149; multivariate analysis, 99–101; partisanship, 92–94, 103, 135–136, 139; policy advocates, 97–98, 102; political implications, 103; pretest interview, 69–70; professionalization, 101–102; proponents, 95, 97, 100, 102; regionalism, 94–95, 135–136; relevant-irrelevant test, 71; routine practice, 75; statistics, 71–72; syndrome evidence comparison, 5, 108–109, 112, 118–120, 122, 126, 128–129, 133, 135; technique, 70–72, 89; theory of, 68–70, 73; third-party support, 98–99; validity, 71–73
polymerase chain reaction analysis, *see* PCR analysis
polymorphisms, 37–39, 43, 74
popular culture, 159–162
population frequency statistics, *see* forensic DNA: match statistics
Posner, Daniel N., 26
post-sentencing hearings, 75
post-traumatic stress disorders, 105–106, 112, 131
power, 119
precedent, 43–45, 100–101, 115
pre-sentencing hearings, 75
President of the United States, 3, 39, 147, 152–154
President's National Advisory Committee for Violence Against Women, 153
pretest interview, 69–70
pre-trial motions, 129
Prichett, C. Herman, 17, 20
probability, 38–39
probation, 75
procedural innovations, 120. *See also* middle ground approach
product rule, 38–39, 61
professional associations, 27, 54, 73, 154
professionalization, 27–28, 39, 88
proof, 106
Prosecuting Attorney's Appellate Service, 97
Prosecuting Attorneys Association, 97

prosecution, 2, 11, 30, 49, 55, 61, 67, 73, 75, 92, 95–97, 106–108, 123–124, 126–128, 133, 136–137, 152, 162; amici briefs, 58–60; prosecutors, 5, 22, 25, 50, 52, 56, 58–59, 61, 73, 93–94, 96–98, 128–129, 139, 144–149, 153, 160–161; strategy, 73. *See also* law enforcement
psyche, 113
psychiatry, 86–87
psychological satisfaction, 91, 102
psychological syndromes, 105–133; nonlegal applications, 108. *See also* battered woman syndrome; rape trauma syndrome; syndrome evidence
psychological tests, 13
psychologists, 106–108, 128
psychology, 69–70, 115
PTSD, *see* post-traumatic stress disorders
public defenders, *see* defense
public policy, 21, 103, 141; theories of, 6, 34, 143–148

Raloff, Janet, 154
rape trauma syndrome, 2, 105–133, 146, 152; admissibility, 5, 109–113; compared to forensic DNA evidence, 9–10; definition of, 105–107; expert testimony, 128–129; scientific reliability, 113; testimony, 106; as therapeutic tool, 108. *See also* syndrome evidence
rape victims, 105–107, 152
Raskin, David C., 72
"reasonable man" standard, 113
regionalism, 5, 22, 52–53, 65–67, 94–95, 100, 122, 125–126, 135–136, 139
regression analysis, *see* multivariate analysis
Reid, Tracy V., 12
relevancy standard, 88
repeat players, 24–25, 28, 34–35, 74, 131–132, 138
Report of President's Commission on the Assassination of President Kennedy, 98–99
Republicans, 16, 21–22, 50–52, 54, 66, 92–94, 122–125, 132–133, 136, 138–139, 150
respiration, 69
restriction-fragment length polymorphism (RFLP) analysis, 37–39, 42–43
Rhode, David W., 20
Rhode Island Supreme Court, 43, 114
rights of the accused, 52–53, 125
Romans, Neil T., 12, 27
Rothenberg, Karen H., 155
Rowland, C. K., 15, 21–22, 24

rules of evidence, 148. *See also* Federal Rules of Evidence (FRE) 702

Sabatier, Paul A., 143–148
Schmitt, Glenn R., 152
Schubert, Glendon, 17, 20
Schutz, Amy T., 3
science, *see* scientific knowledge
science courts, 2–3
scientific ambiguity, 46, 109
scientific community, 39, 74, 113
scientific knowledge, 1, 4, 7–9, 12, 34–35, 67–68, 70, 74–75, 90, 98, 103–104, 126, 133, 141–142, 156, 161; criminal law, 55; political importance, 8–9; political power, 23
scientific literature, 97, 100
scientific reliability, 2, 13, 19, 24, 39–40, 42, 61, 69–75, 77–78, 80–82, 85–86, 90, 95, 99, 101, 106–107, 109, 115, 117–118, 121–122, 131, 133–135, 140, 152–153, 155–156, 160
scrutiny, 71
search and seizure, 25; Fourth Amendment, 155, jurisprudence, 17
Seattle/King County Public Defender's Association, 59
Segal, Jeffery A., 17–18, 20–21, 24
self-defense, 107, 113, 121–122, 129, 135
sentencing, 75
separation of powers, 137
sexual abuse and assault, 106
Shapiro, E. Donald, 26
Sheehan, Reginald S., 21, 24–25
Sickels, Robert, 12
Simon, Herbert, 15
single-decision jurisdictions, 43–45, 82, 114–115
sobriety tests, 13
social workers, 106–107, 128
Songer, Donald R., 12, 18, 21–22, 24–25
South Carolina Supreme Court, 45, 114
South Dakota Supreme Court, 45, 83, 114
Spaeth, Harold J., 18, 20–21
spectrograph, 28
Spencer v. Commonwealth, 384 SE 2d. 785 (1989), 36
stare decisis, 17
state attorneys, *see* attorney general; prosecution
state of mind, 74, 121
state policies, 56
state supreme courts, 4, 9, 138, 140, 155; battered woman syndrome, 111, 132; case load, 11; dissent, 12, 50–51; DNA admissibility, 40–43; forensic DNA, 40–42;

gatekeeping policy, 1, 9, 14; innovation, 156; midwestern, 52–53; northeastern, 52–54, 65–66, 145; organizational interests, 12, 27; partisanship, 49–54, 92; policy correspondence, 27; polygraph admissibility, 77–78, 80–81; polygraphs, 68, 76–82, 102; rape trauma syndrome, 106–107, 110, 132–133; regionalism, 5, 22, 52–54, 65–67, 94–95, 100, 122, 125–126, 131, 135–136, 139, 145; reputation, 11, 155; southern, 22, 52–53, 66, 145; study of, 11–12, western, 52–53, 94, 122, 126, 145

State v. Bible, 858 P.2d 1152, 1180 (Ariz. 1993), 37, 42

State v. Cathon, 846 P.2d 502 (Wash. 1993), 42

State v. Esposito, 670 A.2d 301 (Conn. 1996), 85

State v. Fain, 774 P.2d 252, 257 (Idaho 1989), 70

State v. Harrod, 26 P.3d 492 (Ariz. 2001), 70, 75

State v. Lumley, 977 P.2d 914 (Kan. 1999), 75

State v. Schwartz, 447 NW 2d 422 (Minn. 1989) 36

State v. Shivley, 999 P.2d 952 (Kan. 2000), 71

State v. Stanislawski, 216 N.W.2d 8 (Wis. 1971), 85, 87

statistics, polygraphs, 71–72. See also forensic DNA: match statistics

Stern, Ruth C., 160–161

stipulated evidence, 75, 83, 85–86, 91–92, 96–97, 102–103, 121, 135, 162

stipulation, 75–76, 79, 82, 87, 90–91, 94, 96–97, 102–103, 120, 122, 147

strict legal standard, see legal standards: strict or liberal

subjective interpretation, 68–69, 71, 74

superior court decisions, 27

syndrome evidence, 2, 10, 13, 19, 21, 26, 28, 102, 105–134, 151; admissibility, 109–113; amici briefs, 130–131, 133; appellant status, 126–128; attitudinal factors, 104, 122–126; chronological evidence, 115; chronology, 40–41, 64, 66, 79, 86, 89, 97, 109, 133, 149, 153; controversy, 110–111, 118; cottage industry, 108; detection tool, 118; expert testimony, 108, 118, 128–129; forensic DNA evidence comparison, 9–10, 108, 114, 119–120, 126, 128–129, 131–133; holdings, 112; to inform, 118; institutional factors, 108, 126–132; 136–137; judicial outcomes, 109–113; jurisprudence, 114–115; legal factors, 119–122, 135; media coverage, 159; partisanship, 122–125, 135–136, 139; policy advocates, 130–131; political factors, 126–132; political implications, 133; politics, 108–109; polygraph evidence comparison, 5, 108–109, 112, 118–120, 126, 128–129, 133, 135; proponents, 126–128; psychological, 102, 104; rape trauma syndrome, 110; regionalism, 125–126, 135–136; role of legal facts, 121; scientific vulnerability, 109; as therapeutic tool, 106; third-party reports, 131–132. See also battered woman syndrome; psychological syndromes; rape trauma syndrome

Tabrizi, Susan J., 12
Tarr, G. Alan, 156
Tate, C. Neal, 20–21
television, 159–161
Tennessee Supreme Court, 45, 148
Texas Court of Criminal Appeals, 45, 113–114
therapeutic tool, 106
third-party reports, 31–33, 58, 60, 67, 79, 98–99, 102–103, 131–132, 138, 140, 150–151, 162
third-party support, 6, 28–29, 34, 54, 60, 62, 65–66, 95, 133, 137–138, 140
Thompson, William C., 28
threshold for admissibility, 31–34, 88, 132. See also legal standards
time (chronology), 40–41, 64, 66, 79, 86, 89, 97, 109, 133, 149, 153
Tomasi, Timothy B., 21
Traut, C. A., 18
trial courts, 55, 58, 68, 75, 92, 95, 97
trial lawyers, 49, 91
trials, 72, 74, 113, 129
trust, 99, 141
truthfulness, 71–72, 121

Ulmer, S. Sidney, 20–21
uncertainty, 12, 102, 132, 138
United States Attorney General, 153
United States Census, 65
United States Circuit Courts of Appeal, 155
United States Congress Office of Technology Assessment, 79, 140, 150
United States Constitution, 146
United States Department of Defense, 79, 98–99, 140, 153
United States Department of Energy, 141, 151

United States Department of Justice, 39, 152–153
United States House Committee of Government Operations, 99
United States House of Representatives, 70, 98–99, 104
United States military, 86
United States Supreme Court, 19, 27, 47, 68, 79, 82
United States v. Galbreth, 908 F.Supp. 877 (D.N.M. 1995), 156
United States v. Posado, 57F.3d. 428 (5th Cir. 1995), 19
United States v. Scheffer, 523 U.S. 203 (1998), 79
Utah Supreme Court, 43, 83

validity, *see* scientific reliability
Velona, Jess A., 21
Vermont Supreme Court, 50, 64, 82, 114
victim, 21, 105, 112
Violent Criminal Control and Law Enforcement Act of 1994, 151
Virginia Supreme Court, 4, 11, 36, 40, 45–46, 153
voice prints, 9, 13, 138
Vose, Clement E., 24

Walker, Lenore, 107
Warner, William J., 152
Washington Association of Criminal Defense Lawyers, 130
Washington, DC, 153
Washington Supreme Court, 42, 50, 83, 113–114
"weak" record, 74
Wenner, Lettie McSpadden, 22
West Virginia Supreme Court of Appeals, 43, 114
Whatcom County Public Defense, 59
Wheeler, Stanton, 24
white coats, 1–4, 8–9, 11, 140–142
windows of opportunity, 147
Wisconsin Law Journal, 158
Wisconsin Supreme Court, 43, 82, 85
witnesses, 72–73. *See also* expert testimony
Wyoming Supreme Court, 45, 82, 114

ABOUT THE AUTHOR

Rebecca C. Harris (PhD, Political Science) is an assistant professor of politics at Washington and Lee University in Lexington, Virginia. She is a public-policy scholar with an emphasis on law and science. Her current research includes behavioral genetics in the criminal courtroom, bio-politics, and models of public policy.